BABY MOUNTAIN GORILLA

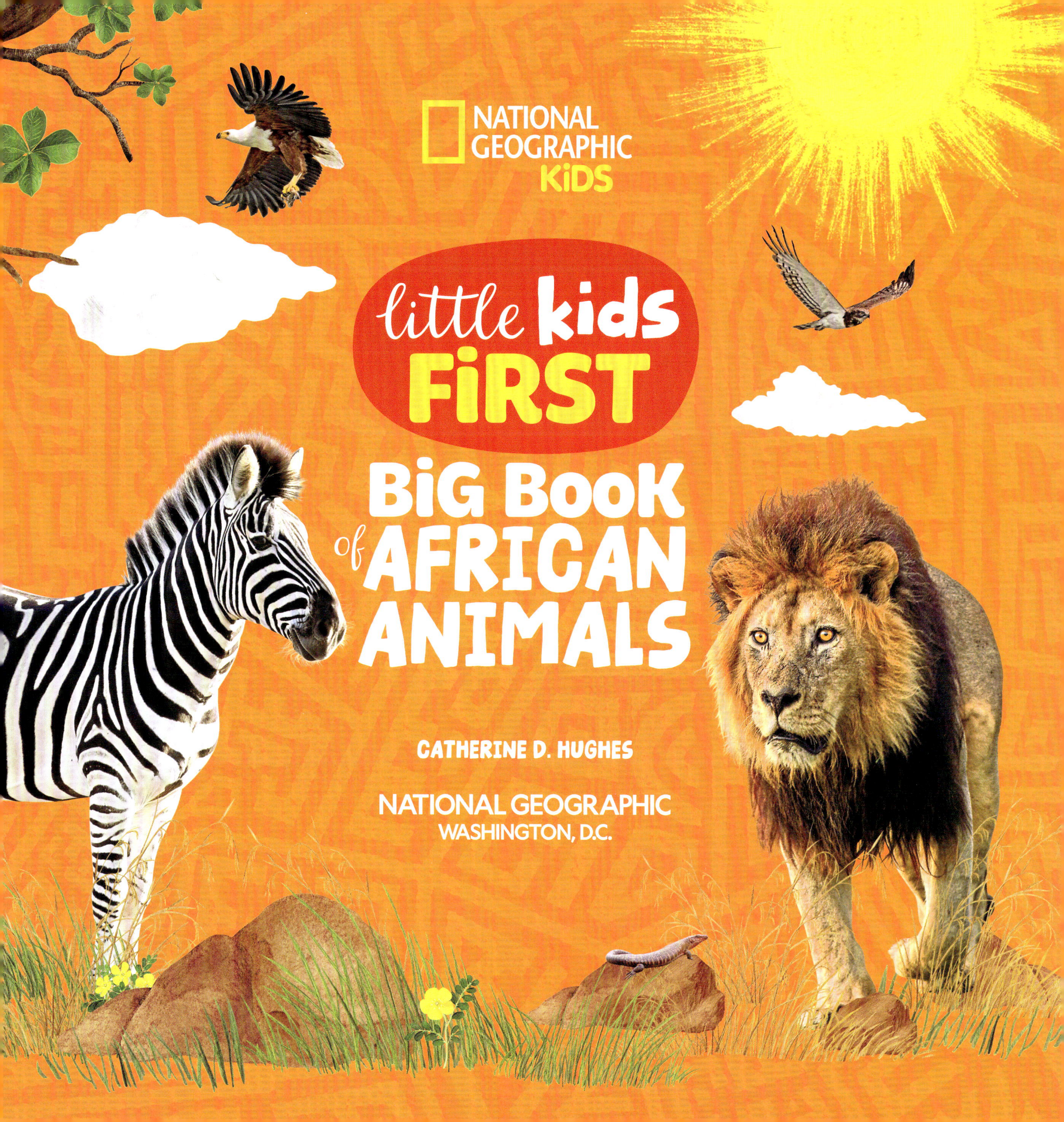

little kids FiRST BiG Book of AFRICAN ANIMALS

CATHERINE D. HUGHES

NATIONAL GEOGRAPHIC
WASHINGTON, D.C.

CONTENTS

INTRODUCTION 6
HOW TO USE THIS BOOK 7

CHAPTER ONE
AFRICA AND ITS WILDLIFE 8
Homes for Wildlife 10
Who Eats What? 12

CHAPTER TWO
POWERFUL HUNTERS 14
Lion 16
African Cat Photo Gallery 18
African Wild Dog 22
Southern African Rock Python 26
African Fish Eagle 28
African Eagle Photo Gallery 30
Nile Crocodile 32

CHAPTER THREE
MORE HUNTERS 36
Aardvark 38
Common Egg-Eating Snake 42
Secretary Bird 44
Banded Rubber Frog 46
Zorilla 48
Bat-Eared Fox 50
King Baboon Tarantula 52

CHAPTER FOUR
PLANTS, PLEASE! 54
Plains Zebra 56
Zebra Photo Gallery 58
Impala 60
African Antelope Photo Gallery 62
African Savanna Elephant 64
Giraffe 66
Wahlberg's Epauletted Fruit Bat 68
Common Warthog 70
Black Rhinoceros 72
African Buffalo 74
Picasso Bug 76
African Insect Photo Gallery 78
Hippopotamus 80

CHAPTER FIVE

ANYTHING GOES 82
Ostrich 84
Chimpanzee 86
African Ape Photo Gallery 88
Giant Plated Lizard 90
Senegal Bushbaby 92
Gray Crowned Crane 94
Diana Monkey 98
African Monkey Photo Gallery 100
Large-Spotted Genet 102

CHAPTER SIX

CLEANUP CREW 104
White-Backed Vulture 106
African Vulture Photo Gallery 108
Brown Hyena 110
Marabou Stork 114
Side-Striped Jackal 116
Dung Beetle 118

MAP OF AFRICA 120
PARENT TIPS 122
GLOSSARY 124
ADDITIONAL RESOURCES 125
INDEX 126
PHOTO CREDITS 127
ACKNOWLEDGMENTS 128

INTRODUCTION

Let's travel to Africa to meet its amazing wildlife! This book introduces dozens of animals that are native to this large, diverse continent. Readers will go on a wildlife adventure to see African mammals, birds, reptiles, amphibians, insects, and more. Each chapter groups animals by what they eat, highlighting carnivores (big and small), herbivores, omnivores, and scavengers. Colorful photo galleries in every chapter cover a wide range of species.

CHAPTER ONE begins with an overview of Africa and its wildlife. More than 50 countries make up Africa, and the continent's habitats range from deserts to rainforests. The animals of Africa are equally diverse.

CHAPTER TWO introduces carnivores that hunt big prey, such as zebras and wildebeests. Readers will meet lions, wild dogs, eagles, and other predators, including the largest snake in Africa.

CHAPTER THREE brings more meat-eaters. The animals in this chapter hunt for small prey, including insects and fish. One featured creature eats only bird eggs!

CHAPTER FOUR explores Africa's herbivores, a diverse group of species that eat plants. Covering grazers and browsers, fruit-eaters and nectar sippers, this chapter's animals range from huge elephants to tiny bugs.

CHAPTER FIVE features the omnivores. A little bit of this, a little bit of that—the diets of these mammals and birds are varied.

CHAPTER SIX wraps up the book with nature's cleanup crew—the scavengers. Mammals, birds, and an insect highlight the importance of this group of animals. They play an essential role in keeping the environment healthy.

HOW TO USE THIS BOOK

COLORFUL PHOTOGRAPHS illustrate each spread and support the text. Several galleries showcase other species related to the featured animal.

POP-UP FACTS provide added information about the animals featured in each section.

PLAINS ZEBRA

Zebra families stick together.

Plains zebras live in small groups called herds. A herd is made up of one adult male, several adult females, and their young. These family groups often gather with other family groups.

FACTS

KIND OF ANIMAL mammal

HOME savannas, woodlands, shrublands in eastern and southern sub-Saharan Africa

SIZE about the height of a pony

FOOD mainly grasses

SOUNDS bark, snort, bray, huff

BABIES one at a time

56

When a predator, such as a lion, threatens a herd, the male zebra makes a loud snort to warn the females and their young. He faces the lion while his family races away.

To escape a lion, cheetah, or other predator, an adult zebra can RUN UP TO 43 MILES AN HOUR (70 km/h).

PLANTS, PLEASE!

If the lion comes close, the zebra stretches out his neck and shows his teeth. He might give the lion a powerful kick. As soon as he can, the male zebra runs away.

What clothes could you wear to LOOK LIKE A ZEBRA? Try it!

57

FACT BOXES for each featured species give readers a quick overview, including kind of animal, range and habitat, size, diet, sounds it makes, and number of young.

INTERACTIVE QUESTIONS encourage conversation related to the animal profile.

The back of the book offers **PARENT TIPS** with fun activities that relate to African animals, along with a helpful **GLOSSARY**.

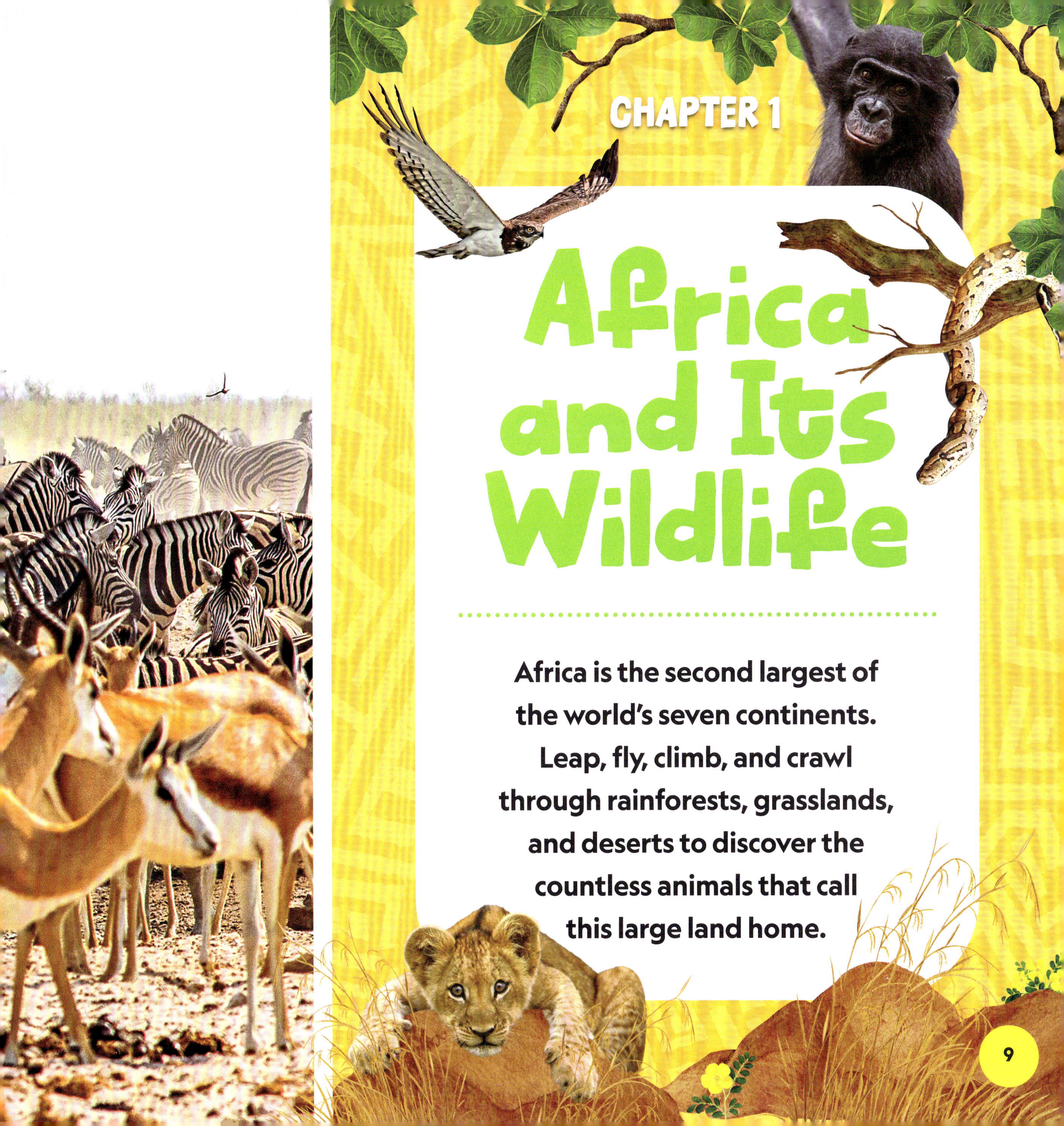

CHAPTER 1

Africa and Its Wildlife

Africa is the second largest of the world's seven continents. Leap, fly, climb, and crawl through rainforests, grasslands, and deserts to discover the countless animals that call this large land home.

HOMES FOR WILDLIFE

Africa has many types of habitats, or places where animals live, including rainforests, deserts, and savannas. More than one million kinds of animals call this huge continent home. Most of Africa is tropical. Tropical areas are warm all year.

A **RAINFOREST** has tall trees that grow close together. They tower over the leafy plants on the forest floor. It rains almost every day in a rainforest!

A **DESERT** has few plants and gets hardly any rain. A huge desert called the Sahara covers a large part of northern Africa. The area south of this desert is called sub-Saharan Africa.

A **SAVANNA** is a grassy area with very few trees. These places have warm weather all year.

EACH ANIMAL SPECIES needs a habitat that provides it with the right kind of shelter and food.

WHO EATS WHAT?

Would a lion chow down on a bunch of grass? Would a zebra enjoy a meal of meat? No, and no. These African animals have very different diets.

Lions, crocodiles, and eagles are **CARNIVORES,** or meat-eaters. These animals are all predators. That means they hunt and eat other animals.

Zebras, elephants, and giraffes are **HERBIVORES,** or plant-eaters. Hungry predators often hunt herbivores. The animals that predators hunt are called prey.

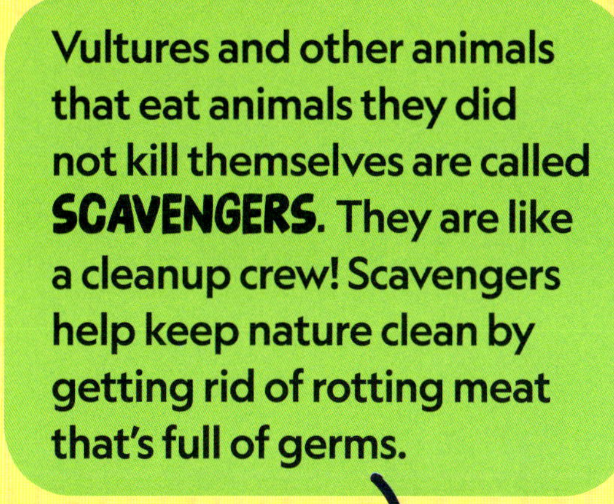

Vultures and other animals that eat animals they did not kill themselves are called **SCAVENGERS.** They are like a cleanup crew! Scavengers help keep nature clean by getting rid of rotting meat that's full of germs.

OMNIVORES, such as ostriches and monkeys, are animals that eat both meat and plants. They can be either predator or prey.

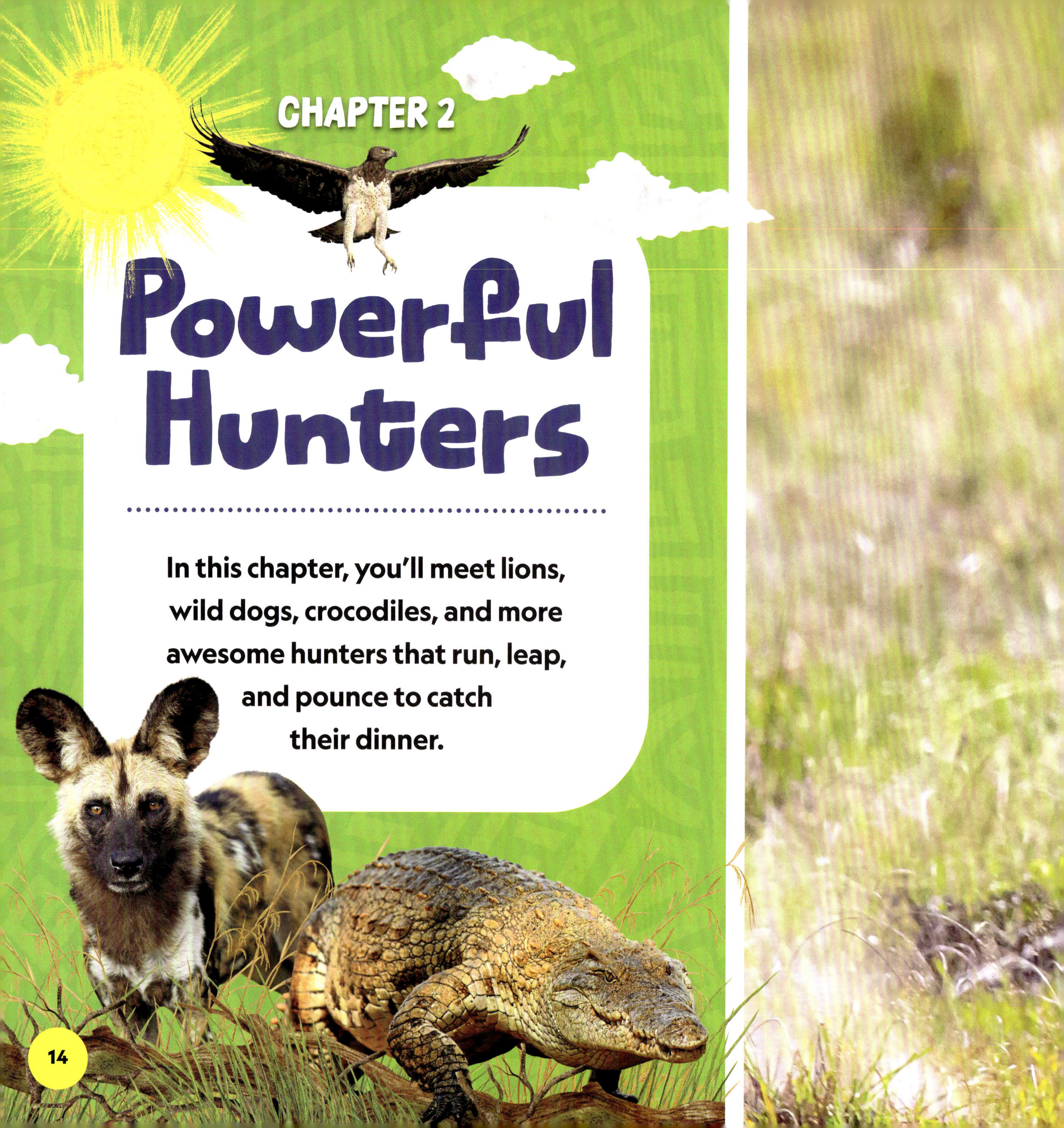

CHAPTER 2

Powerful Hunters

In this chapter, you'll meet lions, wild dogs, crocodiles, and more awesome hunters that run, leap, and pounce to catch their dinner.

LION

Female lions do most of the hunting.

Lionesses, or female lions, take care of the cubs and hunt for food to feed their family, called a pride. Male lions protect the pride's territory from other lions.

FACTS

KIND OF ANIMAL mammal

HOME many habitats in sub-Saharan Africa

SIZE about as long as a twin mattress

FOOD zebras, buffalo, wildebeests, impalas, and other medium and large animals

SOUNDS roar, growl, grunt

BABIES one to six at a time

Lionesses on the hunt usually work together to capture prey. They creep toward the animal they want to chase and spread out to surround it. Then they run toward their prey and bring it down together.

A **PRIDE** usually has about **15 LIONS.**

The pride gathers to eat. Male lions in the pride usually eat first. The lionesses eat next, and the cubs eat last.

HOW MANY PEOPLE live together in your family?

BESIDES THE LION, nine other kinds of wild cats live in Africa. Here are four of them. Turn the page to see the others!

AFRICAN WILDCAT
AFRICAN GOLDEN CAT

SERVAL

Here are **THE OTHER FIVE SPECIES OF WILD CATS** found in Africa.

SAND CAT

BLACK-FOOTED CAT

JUNGLE CAT

CARACAL

AFRICAN WILD DOG

African wild dogs are often called painted dogs.

The furry coat of an African wild dog is white, tan, black, and brown. Each dog's patchy pattern is different from every other dog's coat.

A group of African wild dogs is called a pack. One male and one female lead each pack. The leaders are the only ones in a pack that have pups, but the whole pack helps take care of these babies.

FACTS

KIND OF ANIMAL mammal

HOME deserts, forests, savannas in southern and eastern Africa

SIZE about as big as a standard poodle

FOOD antelope, warthogs, young wildebeests, rodents, birds

SOUNDS bark, howl, twitter, whine, growl

BABIES two to 20 at a time

These **WILD DOGS** are in the same family as pet dogs, gray wolves, jackals, and coyotes.

Wild dogs **EAT THEIR PREY QUICKLY** so that hyenas won't steal their food.

After a hunt, adults eat their prey and carry the food inside their stomachs back to the pups' den. The hunters throw up food for the babies to eat. The pups gobble it up!

Can you **DRAW A PICTURE** using every color in an African wild dog's fur coat?

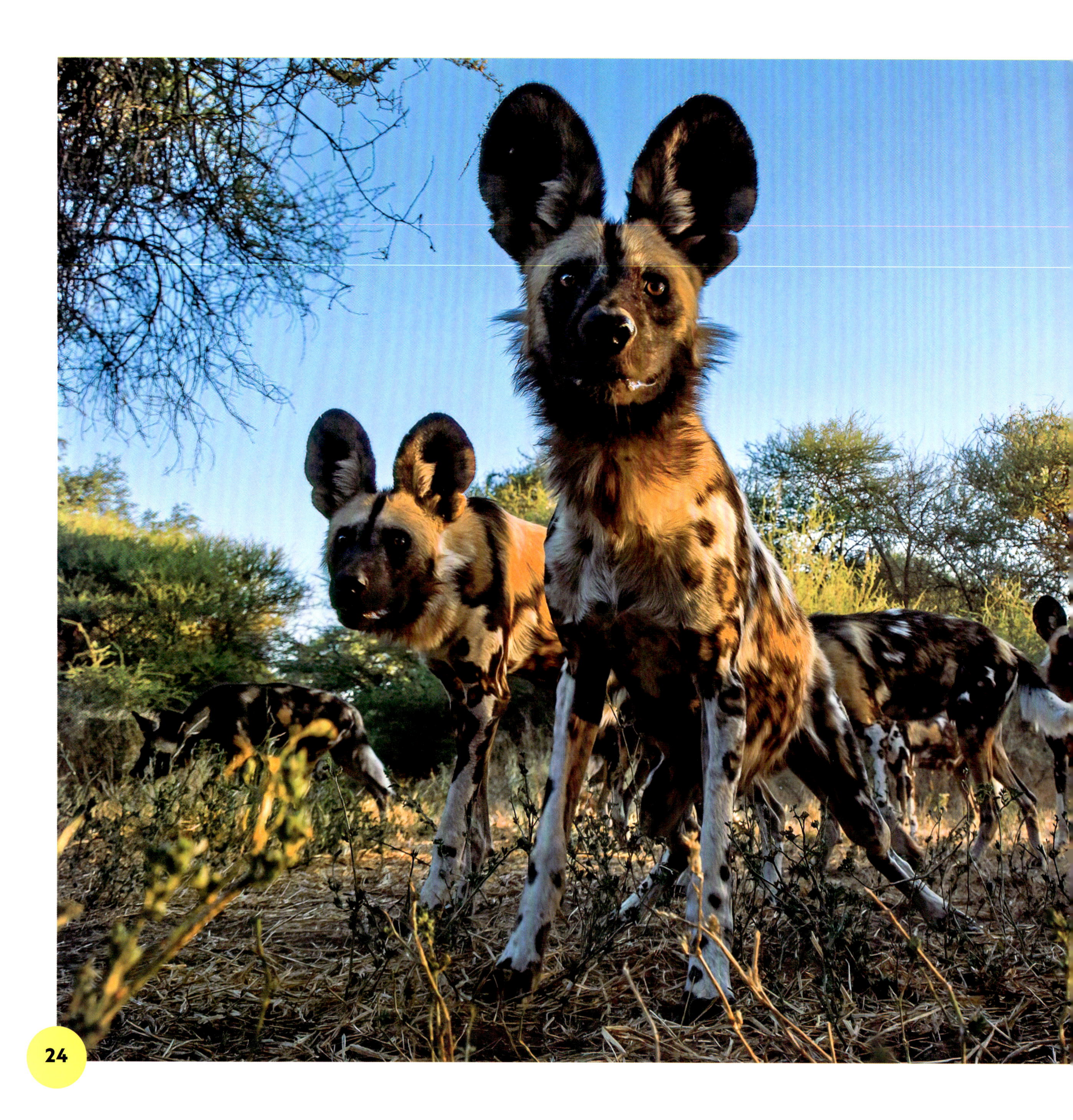

African wild dogs hunt together as a pack. A pack usually has about 10 members. Big packs might have up to 40 wild dogs.

These dogs are great hunters. Once they pick their prey, it rarely escapes. When an African wild dog gets tired of chasing, another pack member takes its place. Together, they wear out their prey until it can't run anymore. Then the wild dogs get to eat.

Wild dogs **TAKE CARE OF SICK OR WEAK PACK MEMBERS** by bringing them food.

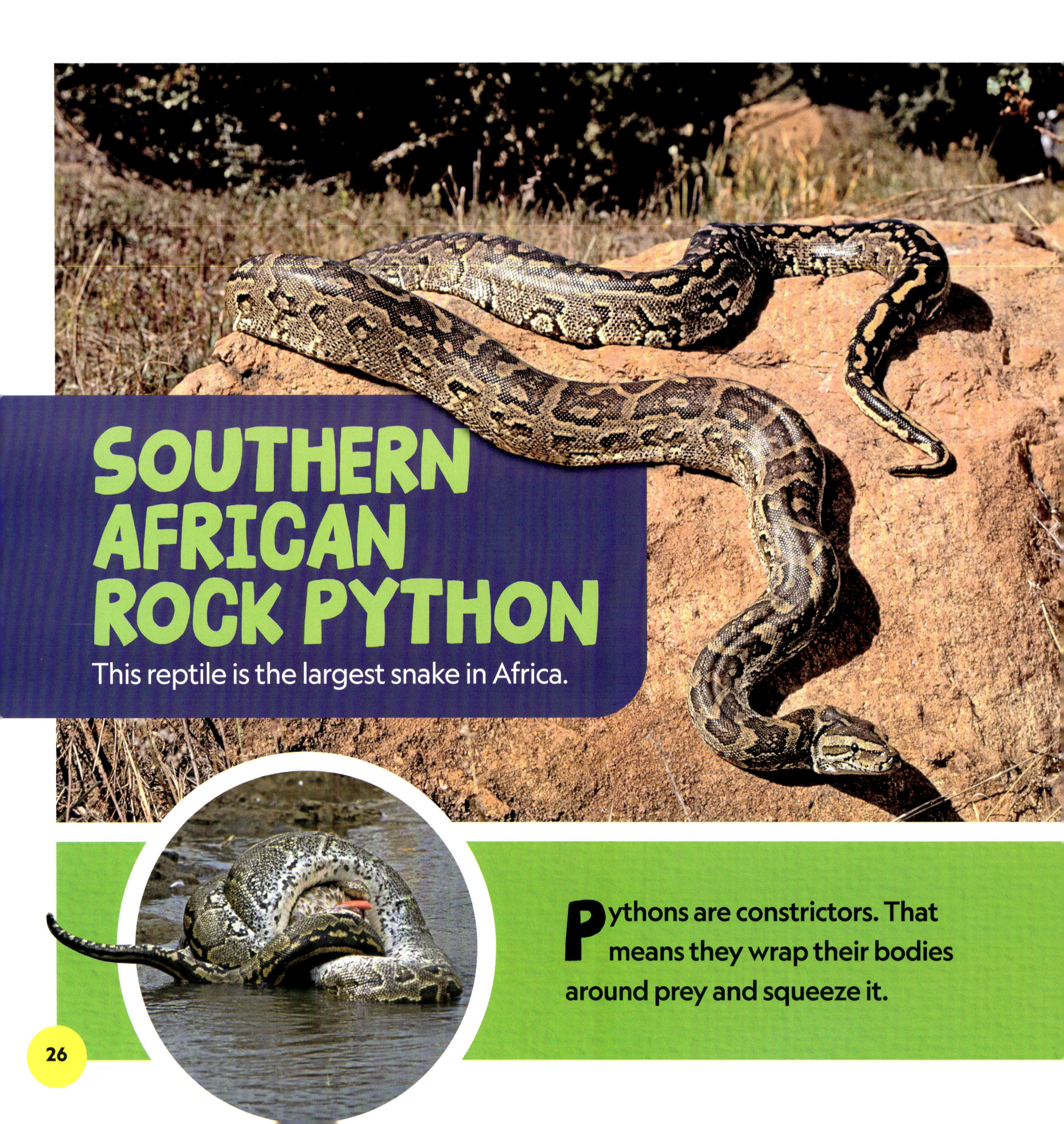

SOUTHERN AFRICAN ROCK PYTHON

This reptile is the largest snake in Africa.

Pythons are constrictors. That means they wrap their bodies around prey and squeeze it.

A Southern African rock python first grabs prey with its long teeth. Then the snake wraps its body around the animal. The constrictor uses its strong muscles to squeeze until its prey cannot breathe. When the prey is dead, the snake swallows its meal whole.

The Southern African rock python is a **GOOD CLIMBER** and **SWIMMER.**

FACTS

KIND OF ANIMAL reptile

HOME most habitats throughout sub-Saharan Africa

SIZE about the length of four four-year-old children lying head to toe

FOOD monkeys, rats, antelope, warthogs, and other animals

SOUNDS hiss

BABIES 20 to 60 eggs at a time

After a large meal, a rock python may not eat again for months.

How many times a day do you eat, **INCLUDING SNACKS?**

AFRICAN FISH EAGLE

This bird catches slippery prey with its feet.

The African fish eagle is related to the **AMERICAN BALD EAGLE.**

The African fish eagle's name tells you what it eats—fish! Sharp spikes on the eagle's feet help it hang on to wet, slick prey. Its strong talons, or claws, also allow the eagle to grip fish. The eagle will usually fly back to a tree to eat the fish it catches.

FISH EAGLES STAY TOGETHER FOR LIFE. Both parents take care of their young.

Some **AFRICAN FISH EAGLE NESTS ARE SO BIG** that an adult human could lie stretched out inside.

Sometimes this strong bird catches prey that is 10 times its size. That would be like a five-year-old trying to pick up two adults! When the eagle catches a huge fish, it uses its wings to paddle to shore, dragging its prey with it. The bird then eats its big feast on the ground.

FACTS

KIND OF ANIMAL bird

HOME near fresh water in much of sub-Saharan Africa

SIZE wingspan the length of a king-size mattress

FOOD mainly fish

SOUNDS loud wail, soft *quock*

BABIES one to four eggs at a time

You have met the AFRICAN FISH EAGLE. Here are FOUR OTHER EAGLES that live in the wild only in Africa.
LONG-CRESTED EAGLE

PHOTO GALLERY
CROWNED EAGLE
MARTIAL EAGLE
AFRICAN HAWK-EAGLE

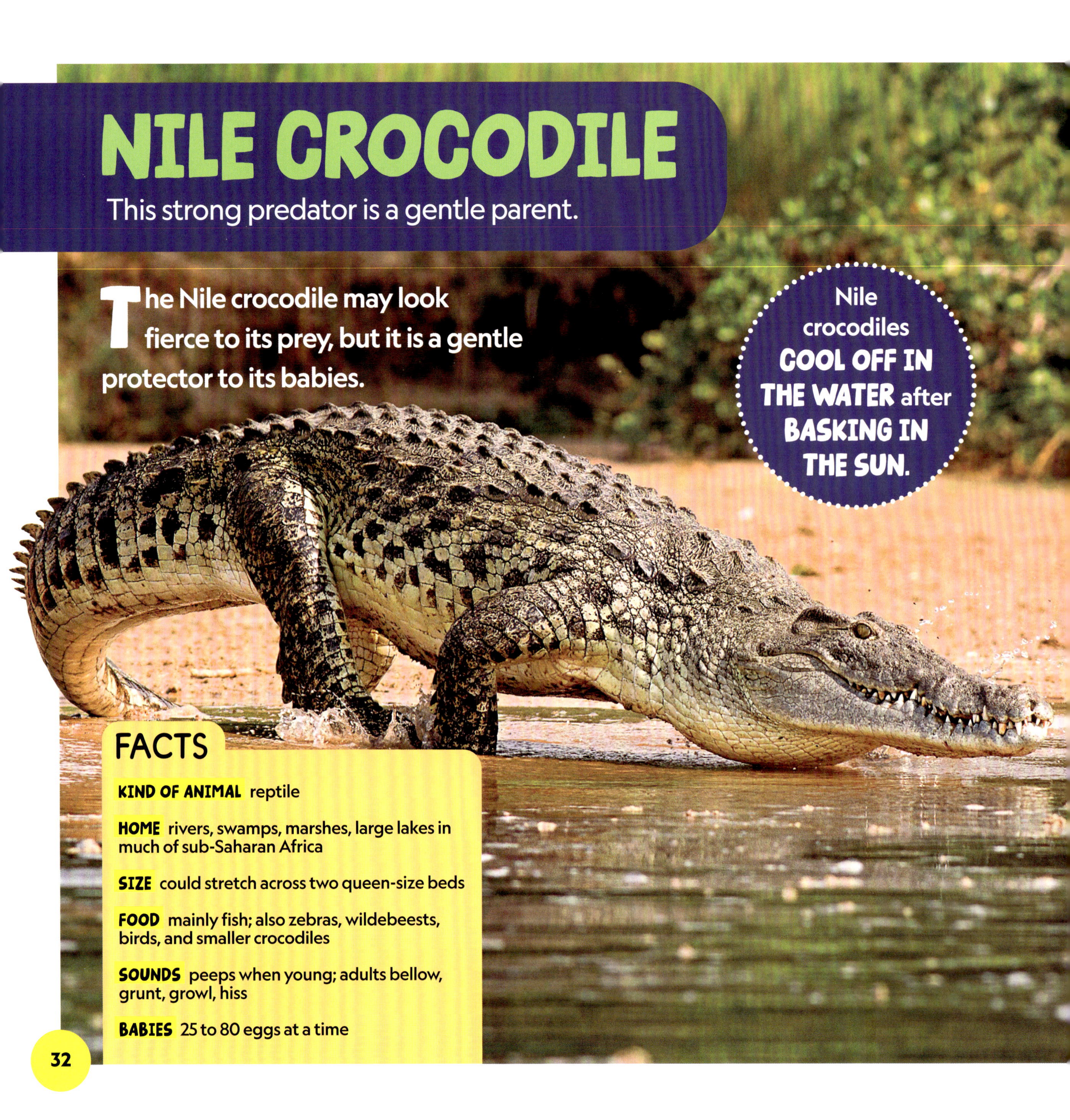

NILE CROCODILE

This strong predator is a gentle parent.

The Nile crocodile may look fierce to its prey, but it is a gentle protector to its babies.

Nile crocodiles **COOL OFF IN THE WATER** after **BASKING IN THE SUN.**

FACTS

KIND OF ANIMAL reptile

HOME rivers, swamps, marshes, large lakes in much of sub-Saharan Africa

SIZE could stretch across two queen-size beds

FOOD mainly fish; also zebras, wildebeests, birds, and smaller crocodiles

SOUNDS peeps when young; adults bellow, grunt, growl, hiss

BABIES 25 to 80 eggs at a time

Both male and female crocodiles guard the eggs in their nest. When it is time for the eggs to hatch, the parents help the tiny babies break out of their shells. The adult carefully rolls an egg in its mouth to crack it. Then the baby can wiggle its way out and crawl into the nest.

BABIES PEEP when they are **READY TO HATCH.**

After the babies hatch, the mother crocodile carries them around in her mouth to the water. Unlike most reptiles, the mother takes care of the babies for a few weeks. A group of baby crocodiles, called a crèche, live together for about two years before they swim away to live on their own.

What kinds of **BABY ANIMALS** have you seen?

Nile crocodiles eat mostly fish, but they grab almost any animal they can. A zebra or other thirsty animal that drinks at the edge of a river might not see a crocodile quietly floating nearby.

A Nile crocodile can **HOLD ITS BREATH UNDERWATER** for two hours.

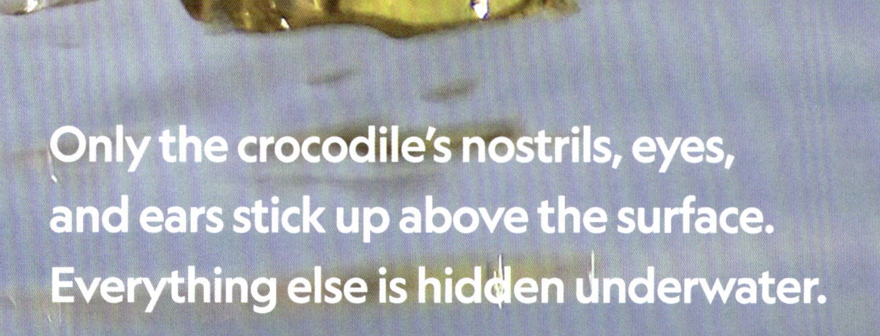

Only the crocodile's nostrils, eyes, and ears stick up above the surface. Everything else is hidden underwater.

As the zebra begins to drink, the crocodile leaps. It snatches the prey in its strong jaws and pulls it underwater.

CHAPTER 3

More Hunters

The animals in this chapter are hunters, too. They prefer small meals, like insects and frogs. You'll also read about one that eats only bird eggs!

AARDVARK

Digging is this mammal's superpower.

An aardvark's front feet have toes with strong, flat nails that look like little shovels. It uses these feet to dig up its food. This animal eats only ants and termites.

The **AARDVARK** gets its name from a South African word meaning **"EARTH PIG."**

An aardvark can gobble up **50,000 ANTS** in one meal.

The termites and ants that aardvarks find tasty live in nests inside dirt mounds. The walls of the mounds are like cement. They are hard to break through.

But aardvarks use their tough nails to easily dig into the mounds. Then they use their long, sticky tongue to lick up thousands of insects inside the nest.

An aardvark's **TONGUE IS ABOUT 12 INCHES** (30 cm) **LONG**. That's about four times as long as yours!

FACTS

KIND OF ANIMAL mammal

HOME savannas, dry areas in much of sub-Saharan Africa

SIZE from nose to tip of tail, the length of a twin-size bed

FOOD ants, termites

SOUNDS grunt, bleat

BABIES one at a time

Aardvarks also use their digging skills to create holes in the ground called burrows. During the heat of the day, aardvarks rest in their cool, dark burrows. After the sun sets, they come out from underground to look for food.

Female aardvarks give birth in burrows. The babies, called cubs, drink their mother's milk. They stay underground until they are a few weeks old. Then they follow their mother as she searches for food. By the time they are six months old, young aardvarks can live on their own.

COMMON EGG-EATING SNAKE

This snake eats one thing only—bird eggs.

After a meal, an **EGG-EATING SNAKE** can go months before eating again.

Common egg-eating snakes are great climbers. They spend most of their time in trees. That is where they can find bird eggs to eat.

FACTS

KIND OF ANIMAL reptile

HOME savannas in much of eastern and southern sub-Saharan Africa

SIZE about as long as a baseball bat

FOOD bird eggs

SOUNDS hissing noise made by rubbing scales together

BABIES six to 28 eggs at a time

The snake wraps its mouth around an egg. Then it opens its jaws extra wide to swallow the egg. An egg-eating snake can swallow an egg that is three times bigger than its head!

THIS SNAKE LIVES where there are a lot of **BIRDS.**

The egg-eating snake has muscles and special bony parts inside its body that break the egg open as the snake swallows it. The snake squeezes out the insides of the egg, then spits out the crushed empty shell.

What is your **FAVORITE WAY TO EAT EGGS?** Fried, boiled, scrambled, some other way, or not at all?

SECRETARY BIRD

This hunting bird doesn't fly much.

The secretary bird is a bird of prey, which means it hunts other animals for food. It is related to eagles and hawks. Most birds of prey hunt as they fly. But secretary birds hunt on the ground.

FACTS

KIND OF ANIMAL bird

HOME savannas, semi-deserts, shrublands in most of sub-Saharan Africa

SIZE wingspan the length of a king-size mattress

FOOD insects, small mammals, birds, eggs, amphibians, reptiles

SOUNDS croaking growl, whistle, squeal

BABIES one to three eggs at a time

Secretary birds **BUILD NESTS** out of sticks in trees. **THEY USE THE SAME NEST FOR YEARS!**

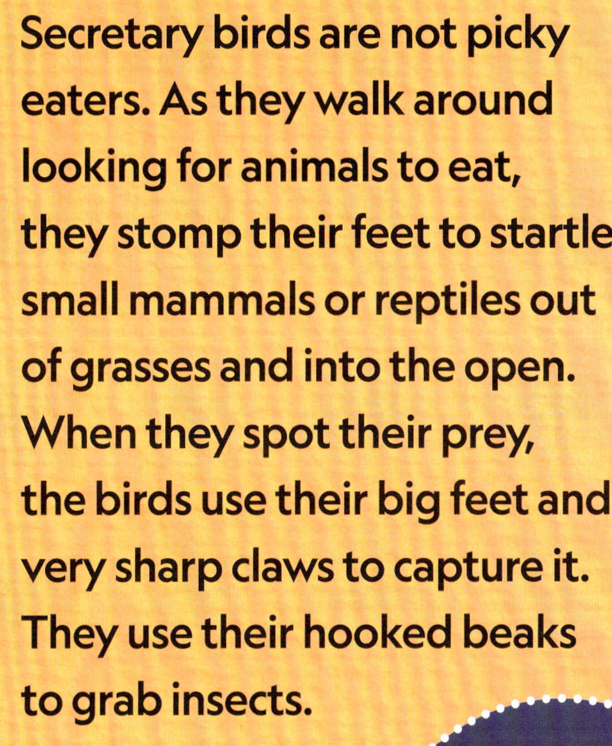

Secretary birds are not picky eaters. As they walk around looking for animals to eat, they stomp their feet to startle small mammals or reptiles out of grasses and into the open. When they spot their prey, the birds use their big feet and very sharp claws to capture it. They use their hooked beaks to grab insects.

TOUGH SCALES on this bird's legs **PROTECT IT** from snakebites.

Are you **TALLER** or **SHORTER** than the four-foot (1.2-m)-tall secretary bird?

BANDED RUBBER FROG

This frog doesn't hop—it walks!

The frog gets its name from the **RUBBERY TEXTURE** of its **SKIN**.

Most frogs hop to get around. But this frog uses all four of its legs to walk. It can even run! A banded rubber frog usually stays on the ground, but it is also a good tree climber.

FACTS

KIND OF ANIMAL amphibian

HOME savannas, shrublands, near fresh water in much of sub-Saharan Africa

SIZE about the width of a credit card

FOOD ants, termites

SOUNDS males trill; females are silent

BABIES about 600 eggs at a time

The frog's colorful bands, or stripes, send a warning to predators: *Stop. Leave me alone!* Its skin can release a toxin, which is a poisonous liquid. This toxin tastes bad and helps keep the frog safe from predators that do not want a yucky-tasting meal.

Banded rubber frog **TADPOLES** hatch four days after the mother frog lays her eggs.

The banded rubber frog digs itself a burrow in the ground to rest during the day. It digs using its back legs, entering the ground rear end first. Sometimes it hides in a hole in a tree.

What **COLOR** would you want to be if you were a frog? **WHY?**

ZORILLA

A zorilla sprays to stay safe.

If a predator gets too close to a zorilla, it better watch out! When threatened, a zorilla turns its back toward the predator, aims, and fires a stinky spray from below its tail. The spray stings and burns the eyes of the attacker. That gives the zorilla a chance to run away.

The zorilla hunts for small creatures at night. It uses the long, sharp claws on its front paws to dig in the dirt for food.

The name **"ZORILLA"** comes from the Spanish word for **"SKUNK."**

FACTS

KIND OF ANIMAL mammal

HOME most dry areas throughout sub-Saharan Africa

SIZE as long as an adult house cat

FOOD small rodents, birds and bird eggs, snakes, amphibians, insects

SOUNDS growl, bark, scream

BABIES one to five at a time

ZORILLAS are also called **STRIPED POLECATS.**

BAT-EARED FOX

Big ears help this fox find food.

A male and female pair usually **STAY TOGETHER FOR LIFE.**

FACTS

KIND OF ANIMAL mammal

HOME hot, dry grasslands and savannas in eastern and southern sub-Saharan Africa

SIZE about the size of a small dog

FOOD mainly termites and beetles; also bird eggs, baby birds, lizards, mice

SOUNDS soft whistle

BABIES two to five at a time

A bat-eared fox listens for its prey. It walks slowly through short grass. The fox keeps its nose close to the ground and its large ears pointed forward.

Termites, the bat-eared fox's favorite food, live underground. But the insects come to the surface to feed on grass. That is when the fox hears them and licks them up off the ground.

Both the **MOTHER** and **FATHER** take care of their **PUPS.**

Bat-eared foxes get most of their **WATER** from the food they eat.

A bat-eared fox eats more than a million termites in a year! A termite is only about the size of an ant, so it takes a lot of them to make a meal.

What is your **FAVORITE FOOD?**

KING BABOON TARANTULA

This is the biggest spider in Africa.

If a king baboon tarantula stretched out on your dinner plate, there would be no room for your food!

King baboon tarantulas **RANGE IN COLOR** from **RUSTY BROWN** to **ORANGE.**

This tarantula uses its extra-long back legs to dig a burrow in dirt. It hides inside the burrow and waits.

When a beetle or other prey wanders by, the tarantula pops out of its burrow. It grabs the beetle with its front legs and sinks its fangs into its prey.

FEMALE king baboon tarantulas are LARGER THAN THE MALES.
FANGS

FACTS

KIND OF ANIMAL arachnid

HOME grasslands, shrub-lands in eastern Africa

SIZE legs could stretch to almost touch the top and bottom of this page

FOOD beetles, cockroaches, other spiders

SOUNDS hissing made by rubbing body parts

BABIES 30 to 180 eggs at a time

Venom—a type of poison—flows from the tarantula's fangs into the beetle. It turns the beetle's insides into liquid. Then the spider slurps up its meal.

Can you **DRAW A PICTURE** of a king baboon tarantula that is big enough to cover a dinner plate?

CHAPTER 4

Plants, Please!

This chapter explores a variety of animals that eat plants. Whether they nibble grass, munch leaves, or sip nectar, these creatures are nature's vegetarians.

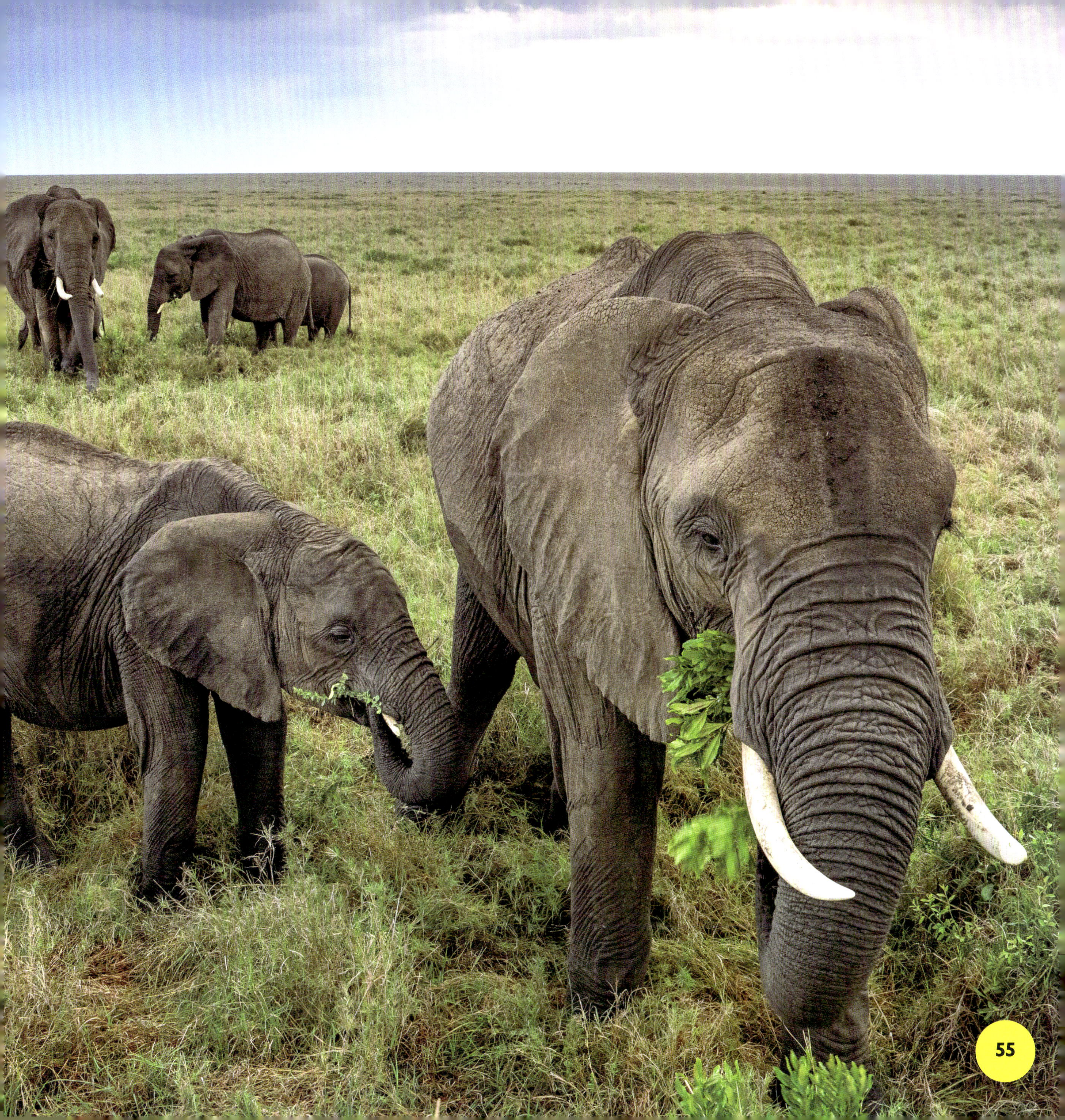

PLAINS ZEBRA

Zebra families stick together.

Plains zebras live in small groups called herds. A herd is made up of one adult male, several adult females, and their young. These family groups often gather with other family groups.

FACTS

KIND OF ANIMAL mammal

HOME savannas, woodlands, shrublands in eastern and southern sub-Saharan Africa

SIZE about the height of a pony

FOOD mainly grasses

SOUNDS bark, snort, bray, huff

BABIES one at a time

When a predator, such as a lion, threatens a herd, the male zebra makes a loud snort to warn the females and their young. He faces the lion while his family races away.

To escape a lion, cheetah, or other predator, an adult zebra can **RUN UP TO 43 MILES AN HOUR** (70 km/h).

If the lion comes close, the zebra stretches out his neck and shows his teeth. He might give the lion a powerful kick. As soon as he can, the male zebra runs away.

What clothes could you wear to **LOOK LIKE A ZEBRA?** Try it!

THERE ARE THREE SPECIES OF ZEBRAS: Grevy's zebras, plains zebras, and mountain zebras. How do these types of zebras look the same? How do they look different?

GREVY'S ZEBRAS are the largest of the zebra species. They live in part of eastern Africa.

PLAINS ZEBRAS are the most common species. They live in grasslands in much of Africa.

MOUNTAIN ZEBRAS use their sharp hooves to dig for water in the dry areas where they live.

IMPALA

This antelope runs and leaps to safety.

Impalas have a few tricks to keep safe from lions, hyenas, and other predators. These antelope live in large herds, so the animals work together to keep watch. If an impala spots danger, it barks an alarm and the whole herd takes off running.

Male impalas have **LONG, POINTED HORNS.** Females **DO NOT HAVE HORNS.**

To get away, impalas run fast in a zigzag pattern. They can cover 33 feet (10 m) in one leap. That's as far as two pickup trucks parked end to end!

When an impala **JUMPS STRAIGHT UP,** it is called **PRONKING.**

Impalas can also jump high. They can leap 10 feet (3 m) into the air. That's like you being able to hop over an elephant!

How far can you **JUMP?**

FACTS

KIND OF ANIMAL mammal

HOME savannas, open woodlands in eastern and southern sub-Saharan Africa

SIZE about the height of a three-year-old child

FOOD grasses, herbs, bushes, shrubs, shoots, fruits, leaves, seedpods

SOUNDS bark, grunt, roar, snort

BABIES one at a time

THERE ARE MORE THAN 80 SPECIES OF ANTELOPE IN AFRICA. Here are just a few of them.

BONTEBOK

GREATER KUDU

BLUE WILDEBEESTS

AFRICAN SAVANNA ELEPHANT

This creature is the largest land animal on Earth.

It takes a lot of food to keep a huge animal like the savanna elephant going. One elephant can easily eat 300 pounds (136 kg) of food in one day. That's about the weight of 15 watermelons!

Elephants have two long tusks, which are giant teeth. They use these curved tusks to dig for roots and tear bark off trees to eat. Elephants also use their tusks to dig for water underground.

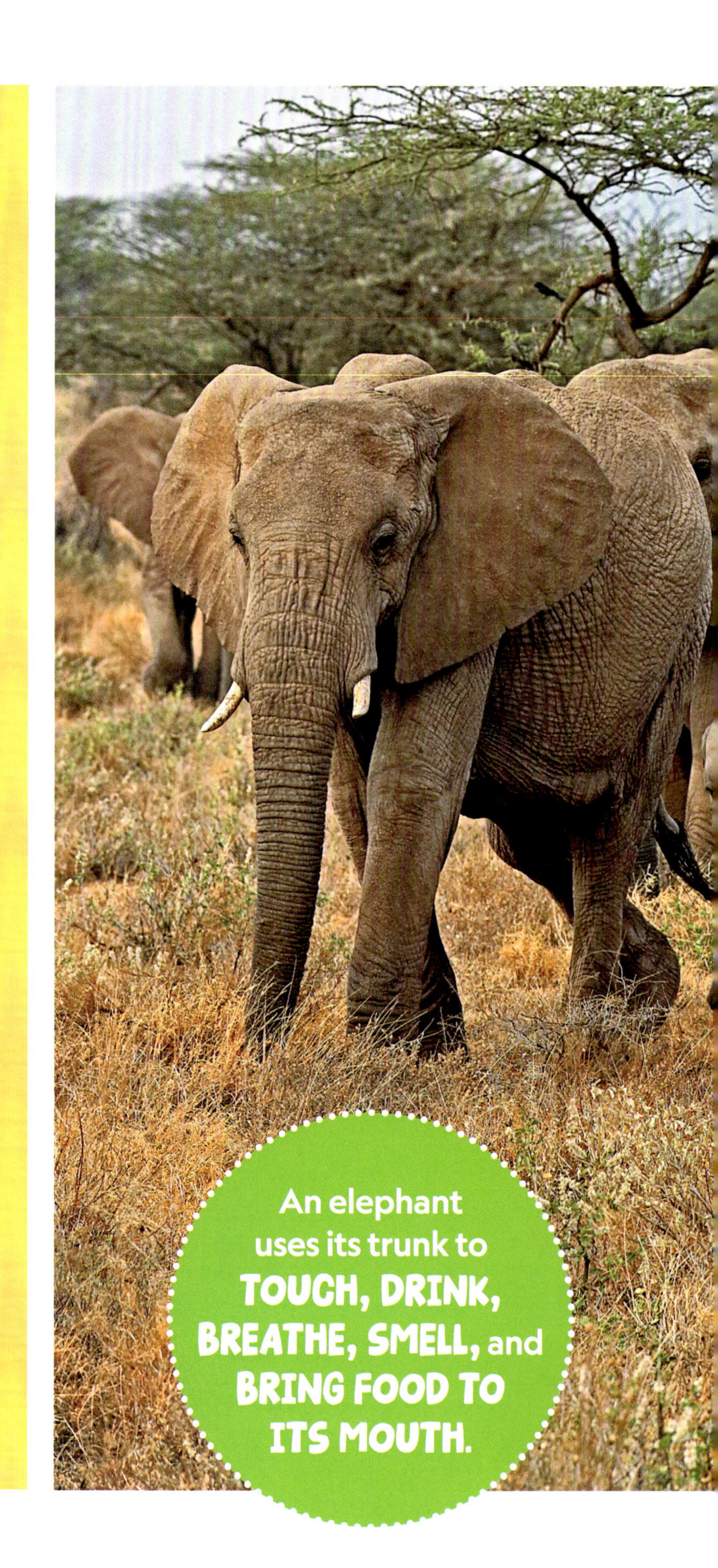

An elephant uses its trunk to **TOUCH, DRINK, BREATHE, SMELL,** and **BRING FOOD TO ITS MOUTH.**

African elephants **FLAP THEIR EARS** to cool off.

FACTS

KIND OF ANIMAL mammal

HOME open and wooded savannas in parts of sub-Saharan Africa

SIZE can weigh as much as three big cars

FOOD grasses, leaves, twigs, roots, fruits

SOUNDS trumpet, rumbling, snort, grunt, bark

BABIES one at a time

GIRAFFE

The giraffe is the world's tallest land animal.

A **GIRAFFE'S TONGUE** could almost stretch across these two pages.

Being tall gives a giraffe an awesome view of everything around it. It has sharp eyesight, too. Giraffes can spot predators, such as lions and leopards, from far away.

FACTS

KIND OF ANIMAL mammal

HOME dry savannas, open woodlands in parts of eastern and southern sub-Saharan Africa

SIZE about as tall as three men standing on each other's shoulders

FOOD leaves, herbs, vines, flowers, fruits

SOUNDS usually silent; grunt, whistle, moan, snore, hiss, bellow, bleat, mew

BABIES one at a time

A giraffe's height also helps it easily reach leaves, fruits, flowers, and other food at the top of trees.

Like a cow, a giraffe chews and re-chews its food. It brings up swallowed food from its stomach to its mouth. This partially digested food is called cud. It may chew cud for hours as it travels looking for more food.

SPOT THE DIFFERENCES

These four kinds of giraffes live in Africa.

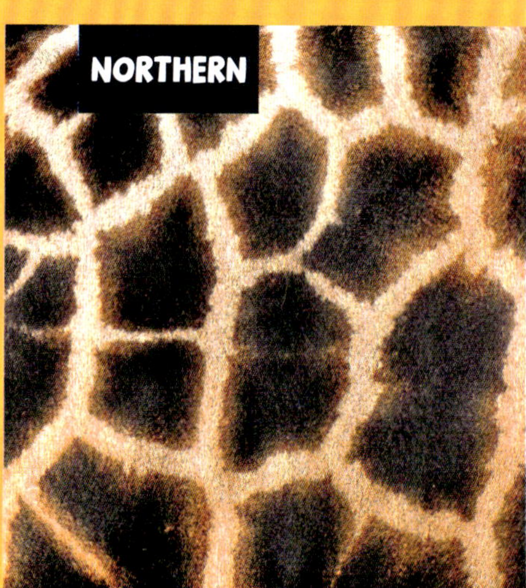

How are the fur patterns above **DIFFERENT** from each other? How are they **ALIKE?**

WAHLBERG'S EPAULETTED FRUIT BAT

This bat helps forests grow.

Wahlberg's epauletted fruit bats mainly eat fruit. They sniff out ripe, ready-to-eat fruit on trees.

These bats are usually active at night. **THEY REST TOGETHER IN GROUPS DURING THE DAY.**

FACTS

KIND OF ANIMAL mammal

HOME savannas, shrublands, forests across southern Africa

SIZE weighs as much as a stick of butter; wingspan as wide as these two pages

FOOD fruits, flowers, leaves, nectar, sometimes insects

SOUNDS chirp

BABIES one at a time

Fruit bats swallow a fruit's seeds, but the seeds do not break down in their stomachs. The bats fly to other areas, carrying the whole seeds inside them. When the bats poop, the seeds come out and fall to the ground. New fruit trees grow from the seeds. These bats also feed on insects and the nectar in flowers.

COMMON WARTHOG

Warthogs belong to the same family of animals as pigs.

The common warthog has two sets of sharp tusks that curve up from its mouth. These tusks make the animal look fierce, but it would rather run from predators than fight them.

WARTHOGS have longer legs than pigs. This helps them **RUN FAST TO ESCAPE** predators.

FACTS

KIND OF ANIMAL mammal

HOME open and wooded savannas, semi-deserts throughout much of sub-Saharan Africa

SIZE about as tall as a one-year-old child; weighs about as much as a big dog

FOOD grasses, roots, bulbs, berries, young tree bark

SOUNDS grunt, chirp, growl, snort, squeal

BABIES two to four at a time

A warthog uses its snout and feet to dig for roots and other buried food. It often bends its front feet back and moves on its wrists while it eats.

WARTHOGS hear and smell very well, but their **EYESIGHT IS NOT GOOD.**

Can you name three things you **HEAR, SMELL,** or **SEE** every day?

These animals are called warthogs because the bumps on their faces look a bit like warts. Male warthogs have bigger bumps than females do. The bumps help protect their eyes and cushion their heads when they fight each other over mates.

BLACK RHINOCEROS

This rhino's lip helps it reach food.

Adult rhinos are **SO BIG** that most **PREDATORS LEAVE THEM ALONE.**

FACTS

KIND OF ANIMAL mammal

HOME deserts, savannas, woodlands, wetlands in parts of eastern and southern Africa

SIZE weighs about as much as a small truck

FOOD leaves, branches, shoots, woody bushes, fruits

SOUNDS grunt, snort, pant, whine, squeak, snarl

BABIES one at a time

The black rhinoceros has a V-shaped, pointed upper lip. This lip helps it eat leaves and fruits on trees and bushes. The rhino reaches up and uses its hooked upper lip to pick food off branches.

A black rhino has two horns on the front part of its head. The front horn is the longer of the two. Males use their horns to fight each other for territory. Females mainly use their horns to defend their babies from predators.

Can you make **EACH OF THE SOUNDS** that a rhino makes?

WHITE RHINO

The white rhino is the only other species of African rhino. Both black and white rhinos are actually grayish in color. The main way to tell them apart is by looking at their lips.

The white rhinoceros has a wide, square upper lip. White rhinos use their broad lips to scoop grasses into their mouths.

AFRICAN BUFFALO

These animals know how to keep their herd safe.

African buffalo are careful. Their herd stays alert. If they smell, hear, or see a lion or other predator, they all stand still. They watch where the predator goes. They may stampede, which means the whole herd runs away together. If they must, they use their horns to fight.

Both **MALE** and **FEMALE** African buffalo **HAVE HORNS.**

FACTS

KIND OF ANIMAL mammal

HOME savannas, forests throughout much of sub-Saharan Africa

SIZE weighs about as much as seven humans

FOOD mostly grasses

SOUNDS bleat, grunt

BABIES one at a time

One
HERD can
include more than
1,000 AFRICAN
BUFFALO!

The whole herd helps mother buffalo protect the babies in the group. A mother buffalo takes good care of her calf. She feeds, cuddles, and plays with her baby. If a calf loses its mother, other females in the herd will take care of it.

Can you think of three ways your family **TAKES CARE OF YOU?**

PICASSO BUG

This colorful insect is named after the famous artist Pablo Picasso.

The Picasso bug's **BRIGHT COLORS WARN PREDATORS:** *I do not taste good!*

A Picasso bug is tiny and colorful. One might look a little bit different from another, but they all have exactly 11 spots.

FACTS

KIND OF ANIMAL insect

HOME dry areas in much of Africa

SIZE would fit on the tip of your finger

FOOD plant juices, nectar

SOUNDS none known

BABIES 30 to 50 eggs at a time

This bug is part of a group of insects called jewel bugs. All insects have three main body parts: the head, the thorax, and the abdomen. Jewel bugs, also called metallic shield bugs, have a thorax that extends to cover their abdomens and wings like shields.

This insect releases a **STINKY SMELL** if something bothers it.

Can you count **11 SPOTS** on the Picasso bug?

A Picasso bug drinks all its food. Its mouth is shaped like a beak. To eat, it sticks its beak into a plant. Its saliva turns bits of the plant into liquid. Then the bug sucks up the liquid through its beak.

THE PICASSO BUG is one of more than 100,000 species of insects found in Africa. Here are a few of Africa's many colorful insects.

PHOTO GALLERY
RAINBOW SHIELD BUG
RAINBOW MILKWEED LOCUST
PURPLE JEWEL BEETLE

HIPPOPOTAMUS

A hippo makes its own kind of sunscreen.

A hippopotamus is mostly hairless, so its skin needs protection from the sun. A hippo has special glands that make a thick, oily, reddish liquid that oozes through little openings in its skin. This liquid keeps it from getting sunburned.

OXPECKER BIRDS eat insects off the hippo's back. This **PROTECTS THE HIPPO** and **GIVES THE BIRDS A MEAL!**

These big animals spend all day in rivers or lakes to stay cool. Their eyes, ears, and nose are all at the top of their head. This allows them to breathe, see, and hear while they soak in the water.

When the sun sets and it gets cooler, hippos leave the water to eat. They walk to grassy areas, where they graze all night. At dawn, they return to the water.

FACTS

KIND OF ANIMAL mammal

HOME rivers, lakes, swamps close to grassy areas in much of sub-Saharan Africa

SIZE weighs as much as 40 adult men

FOOD grasses, shoots, reeds

SOUNDS honk, roar, grunt, chuff

BABIES one at a time

CHAPTER 5

Anything Goes

Animals that eat both meat and plants are called omnivores. In this chapter, you'll meet animals that have a lot of variety in their diets.

OSTRICH

This is the world's biggest bird.

An ostrich is about nine feet (2.7 m) tall. You would need to stand on a grown-up's shoulders to look this bird in the eye!

FACTS

KIND OF ANIMAL bird

HOME dry savanna, woodland in central and southern Africa

SIZE weighs about as much as an adult human

FOOD grass, seeds, leaves, flowers, fruit, roots, insects, lizards, carrion

SOUNDS hiss, chirp, honk, grunt, boom

BABIES up to 11 eggs at a time

An ostrich does not fly, but it can run fast. One could easily keep up with a car driving through a neighborhood. An ostrich also uses its strong legs to defend itself. Its powerful kick can kill a lion or other predator.

THIS OSTRICH IS HIDING!

Ostriches live in **GROUPS CALLED HERDS.**

When an ostrich needs to hide from a predator, this big bird can make itself small. It lies down and stretches its long neck flat on the ground. From a distance, its body looks like a termite mound or a pile of dirt.

What other animals in this book **LIVE IN HERDS?**

CHIMPANZEE

Chimps laugh, play, and can make tools.

Chimpanzees are a lot like humans. For example, they are ticklish in the same places that many people are, and they laugh when other chimps tickle them on the belly or under the arms. They are one of very few animals that laugh!

Chimps play in many of the same ways that human children do, too. They wrestle, run, jump, chase, and swing.

A chimp's **ARMS** are **LONGER THAN ITS LEGS.**

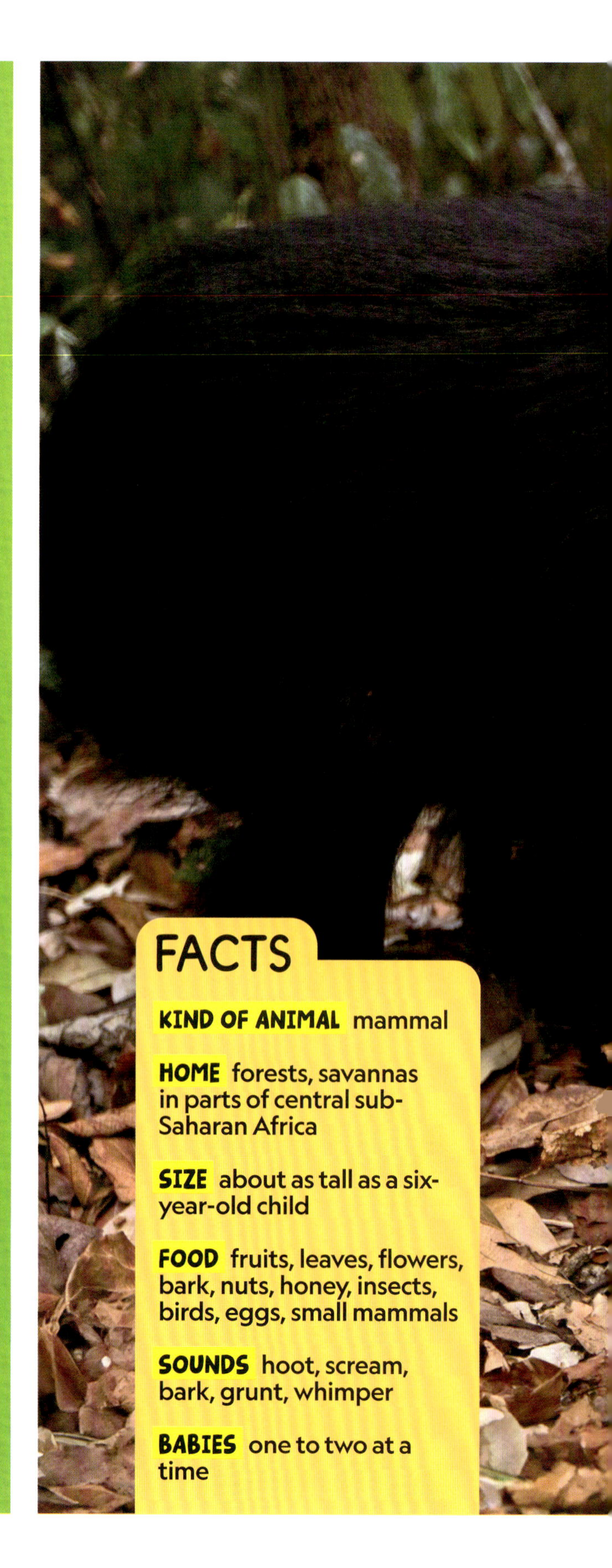

FACTS

KIND OF ANIMAL mammal

HOME forests, savannas in parts of central sub-Saharan Africa

SIZE about as tall as a six-year-old child

FOOD fruits, leaves, flowers, bark, nuts, honey, insects, birds, eggs, small mammals

SOUNDS hoot, scream, bark, grunt, whimper

BABIES one to two at a time

CHIMPS are **NOT MONKEYS.** They are a **SPECIES OF APE.**

Chimps make tools to help them get food. They use grass and twigs to fish termites and ants out of their nests. They use rocks to crack open nutshells. They chew one end of a stick and dip it into a beehive to pick up honey. Young chimps watch adults to learn how to make and use tools on their own.

Where are you most **TICKLISH?**

CHIMPANZEES are one of four species of apes in Africa. The others are the bonobo, the eastern gorilla, and the western gorilla.

PHOTO GALLERY
WESTERN LOWLAND GORILLAS
(A KIND OF WESTERN GORILLA)

GIANT PLATED LIZARD

This lizard has a tricky way to protect itself.

Most of a giant plated lizard's body is covered with large and hard scales in the shape of shields.

FACTS

KIND OF ANIMAL reptile

HOME savannas near rocky hills in much of southern Africa

SIZE up to the length of three copies of this book placed side by side

FOOD flowers, fruits, leaves, insects, small lizards, spiders

SOUNDS none known

BABIES two to five eggs at a time

RUBBERY FEET help giant plated lizards **RUN AROUND ON ROCKS.**

There's a groove, or line, of skin without scales along the lizard's sides. This groove allows the lizard's body to expand, or blow up, like a little balloon.

These lizards stick **CLOSE TO ROCKS,** where they can **HIDE.**

Giant plated lizards have **THICK BODIES** and **STRONG LEGS.**

When it feels threatened, the lizard races toward a rocky area nearby. It squeezes into a crack in the rocks. Once inside the crack, the lizard takes a deep breath. Its body expands. Now the lizard is stuck so tightly in the rocks that the predator cannot pry it out. Tricky!

Where would you go if you were playing **HIDE-AND-SEEK?**

SENEGAL BUSHBABY

A bushbaby finds food in the dark.

When the sun goes down, the Senegal bushbaby starts looking and listening for a meal. Its big eyes help this nocturnal animal see well in the dark.

Each of the bushbaby's ears can move independently, or one at a time. That helps the bushbaby focus in on the sound of insects it wants to catch.

BUSHBABIES are also known as **GALAGOS.**

Senegal bushbabies live mainly in trees. They pee on their hands and feet. Damp feet and hands may help bushbabies hold on to branches as they climb and jump through the trees. Also, the smell left behind on branches lets other bushbabies know who is nearby.

A **BUSHBABY** can **MOVE QUICKLY** enough to catch **FLYING INSECTS.**

FACTS

KIND OF ANIMAL mammal

HOME dry forests, savannas, bushlands across central sub-Saharan Africa and a few islands

SIZE about as long as a U.S. dollar bill

FOOD grasshoppers, small birds, eggs, seeds, fruits, flowers

SOUNDS chirp, cry

BABIES one to two at a time

GRAY CROWNED CRANE

This bird has a crown of stiff golden feathers around its head.

Once a male and female gray crowned crane choose each other as mates, they stay together for the rest of their lives.

FACTS

KIND OF ANIMAL bird

HOME a mix of wetlands and savannas in eastern and southern Africa

SIZE about as tall as a four-year-old child

FOOD grasses, seeds, toads, frogs, crabs, lizards, worms, insects, other small animals

SOUNDS honk, peep, purr

BABIES two to four eggs at a time

Each year during the rainy season, the pair performs a special dance. They bob, bow, and jump, showing off their colorful feathers, long wings, and strong legs.

The **GRAY CROWNED CRANE** is one of **15 SPECIES OF CRANES** around the world.

If you could have a **CROWN,** what would it look like?

CRANE CHICKS SWIM to explore around their nests, but ADULTS AVOID SWIMMING.
PARENT ON CRANE NEST

The gray crowned crane couple builds a nest together. Their nest is usually on the ground, hidden in tall plants at the water's edge. The female lays eggs in the nest. Both parents care for the eggs and the young when they hatch.

The crane chicks grow up fast. They can swim 12 hours after they hatch. When the chicks are two days old, they can follow their parents as they search for food. By the time they are about three months old, the gray crowned crane chicks are on their own.

DIANA MONKEY

This monkey uses different alarm calls for different predators.

FACTS

KIND OF ANIMAL mammal

HOME thick forests in western sub-Saharan Africa

SIZE as tall as a one-year-old baby

FOOD fruits, flowers, leaves, insects

SOUNDS many different calls, including an alarm scream

BABIES one at a time

A Diana monkey on the lookout for predators lets other monkeys know when it spots one. It has a call that warns "leopard!" and another call for "eagle!"

The calls tell other monkeys which danger to look for, so they can find the best place to run and hide. The calls also warn the predator that it has been spotted, which might make it give up the hunt.

Diana monkeys have **WHITE BEARDS.**

DIANA MONKEYS live in **GROUPS** of up to **30 MONKEYS.**

Diana monkeys live in big trees in thick forests. They do not come down to the ground very often. During the day, they look in the trees for fruits and insects to eat. At night, they climb up high to sleep in the treetops.

Can you **MAKE UP YOUR OWN DIFFERENT CALLS** that mean "leopard" and "eagle"?

There are more than 160 **SPECIES OF MONKEYS IN AFRICA.** Here are a few of them.

PHOTO GALLERY
COLLARED MANGABEY
MANDRILL

LARGE-SPOTTED GENET

This little mammal acts a lot like a cat.

When a **GENET** needs to see over tall plants, it **STANDS UP** on its hind legs and uses its **TAIL** to **BALANCE.**

FACTS

KIND OF ANIMAL mammal

HOME near water in savannas, woodlands, thick vegetation in parts of southern Africa

SIZE stands about as tall as this book

FOOD rats, mice, birds, snakes, amphibians, insects, fruits, seeds, leaves, grasses

SOUNDS hiss, growl, purr, meow, churr, yap

BABIES one to five at a time

The large-spotted genet looks a bit like a cat. It purrs, meows, and hisses like cats do. It has claws it can pull in, just like a cat. But it is not in the cat family. It is more closely related to the mongoose.

Genets spend a lot of time in trees. Picture jumping from one end of a small car to the other in one leap. That is how far a large-spotted genet can jump between trees.

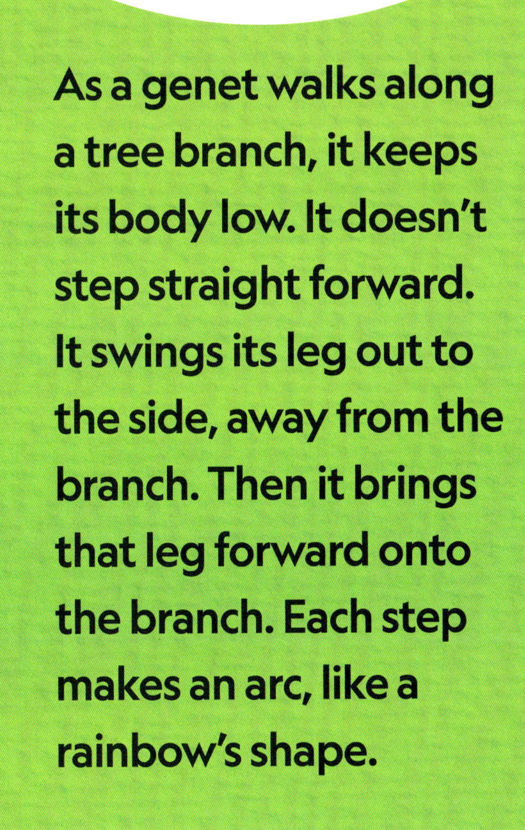

As a genet walks along a tree branch, it keeps its body low. It doesn't step straight forward. It swings its leg out to the side, away from the branch. Then it brings that leg forward onto the branch. Each step makes an arc, like a rainbow's shape.

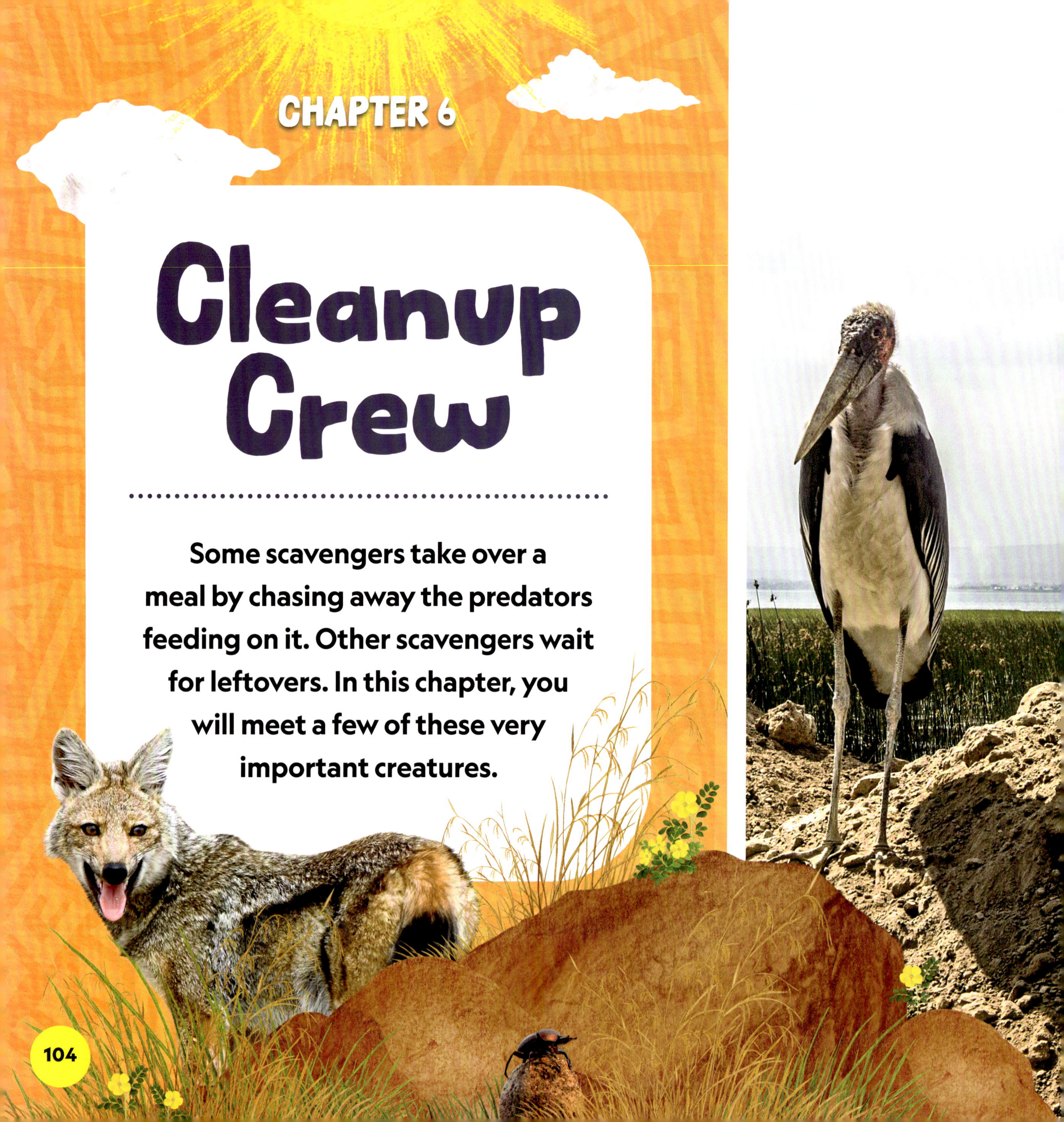

CHAPTER 6

Cleanup Crew

Some scavengers take over a meal by chasing away the predators feeding on it. Other scavengers wait for leftovers. In this chapter, you will meet a few of these very important creatures.

WHITE-BACKED VULTURE

This large bird is the most common vulture in Africa.

White-backed vultures soar around in large circles high in the sky as they look for dead animals, called carrion. They use their sharp eyesight to search for animals that were recently killed by predators.

A group of **VULTURES IN FLIGHT** is called a **KETTLE.**

When one vulture spots a meal, it starts making smaller and smaller circles in the sky. That signals to other vultures that food is below. Dozens of white-backed vultures can gather together to feed on the body of a large dead animal, such as a wildebeest.

How do you let people know when you're **HUNGRY?**

White-backed vultures **NEST IN TREES.**

White-backed vultures **VISIT WATER HOLES** to get **CLEAN AFTER A MESSY MEAL.**

FACTS

KIND OF ANIMAL bird

HOME savannas, forests in much of sub-Saharan Africa

SIZE wingspan the length of a king-size mattress

FOOD carcasses of animals such as warthogs, zebras, gazelles, ostriches

SOUNDS usually silent; squeal, chitter

BABIES one egg at a time

Vultures clean up more than half of all the dead animal bodies, or carcasses, in Africa. The white-backed vulture is one of **11 SPECIES OF VULTURES IN AFRICA.** Here are a few others.

PHOTO GALLERY
CAPE VULTURE
BEARDED VULTURE
LAPPET-FACED VULTURES

BROWN HYENA

These hyenas often steal food.

Brown hyenas are not good hunters. Almost all their food comes from prey killed by larger predators, such as lions and cheetahs. Hyenas have an excellent sense of smell and can find a carcass from several miles away.

FACTS

KIND OF ANIMAL mammal

HOME deserts, semi-deserts, open woodlands, savannas in southern Africa

SIZE about as tall as a two-year-old child

FOOD carrion, including antelope, zebras, jackals; also birds, rodents, insects, eggs, fruits, fungi

SOUNDS whine, squeal, growl, shriek

BABIES one to five at a time

These scavengers are fierce. They race toward jackals, leopards, and cheetahs to chase them away from the prey they are eating. Then the hyenas dig in to eat their stolen meal. Their huge teeth are made for crushing bones.

Brown hyenas are **ACTIVE AT NIGHT** when it is cooler.

Brown hyenas live in desert areas where water can be scarce. They eat melons and other fruits that have a lot of water in them. The fruit gives the hyenas the water they need.

Can you name a food that would help you feel **LESS THIRSTY?**

Brown hyenas live in family groups called clans. They shelter together in one den.

A mother hyena gives birth in a separate den, where she stays with her cubs until they are about four months old.

All the adults in a clan help take care of cubs. They bring meat back to the young hyenas.

A brown hyena **MARKS AREAS WITH A SCENT** that tells others in its clan that **IT HAS SEARCHED THE MARKED AREA FOR FOOD.**

CUBS STAY CLOSE TO THE DEN until they are about **15 MONTHS OLD.**

MARABOU STORK

These birds often live near people.

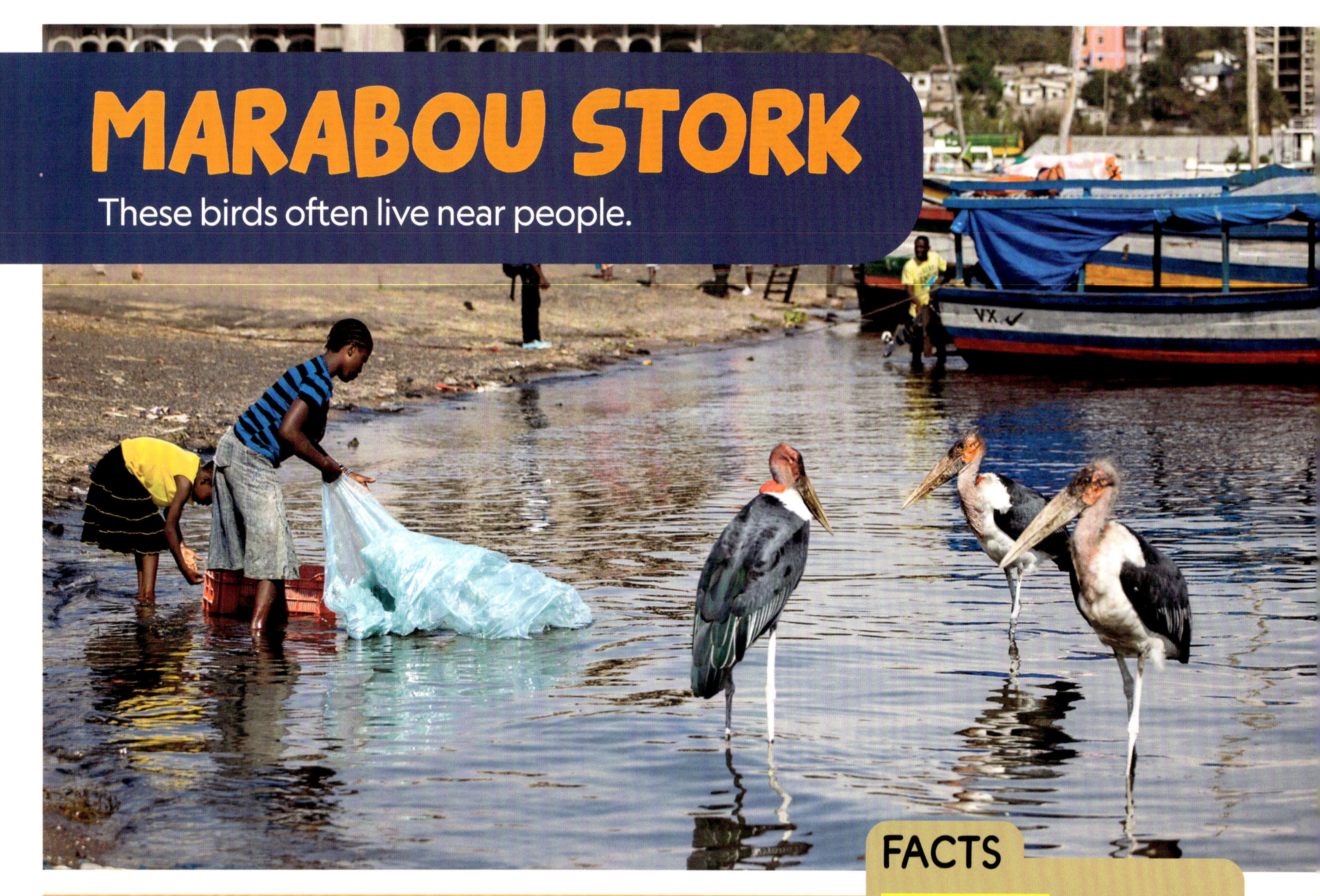

Like other scavengers, marabou storks help prevent the spread of disease by cleaning up carcasses that predators leave. They help clean up garbage left by people, too. The birds sometimes wander streets of villages and cities, looking through trash for food.

FACTS

KIND OF ANIMAL bird

HOME deserts, savannas, wetlands in parts of sub-Saharan Africa

SIZE wingspan is nearly the length of a small car

FOOD mainly carrion; also fish, reptiles, locusts; feeds its young frogs, fish

SOUNDS usually silent; grunt, croak, bill-rattling

BABIES two to three eggs at a time

A marabou stork's head is bald. That way, when it reaches into a messy meal it doesn't get anything stuck in its feathers.

At a feeding site, marabou storks often hang out with vultures. The storks steal the food that the vultures have torn away from the carcass.

STORK STEALING FOOD FROM VULTURES

Sometimes storks WASH THEIR FOOD in water to clean it.

What is the **MESSIEST MEAL** you've ever eaten?

SIDE-STRIPED JACKAL

These jackals find food by following predators.

Side-striped jackals often grab a meal from lions' leftovers. Lions usually ignore the little jackals.

FACTS

KIND OF ANIMAL mammal

HOME wooded areas, savannas, marshes in central and southern Africa

SIZE about the size of a small dog

FOOD carrion, insects, small rodents, reptiles, amphibians, birds, fruits

SOUNDS yip, scream, croak, hoot

BABIES three to six at a time

When a mother jackal is nursing her young, she stays in the den with them. The father jackal swallows big chunks of meat to bring back to the den to feed her. He regurgitates, or throws up, the meat he swallowed so that she can eat it.

Side-striped jackals **USUALLY LIVE ALONE OR IN PAIRS.**

JACKAL PUPS STAY WITH THEIR PARENTS until they are about **11 MONTHS OLD.**

Jackals can be noisy. Family members call to each other with special yips. A jackal facing danger makes a loud screaming sound. Most often, side-striped jackals make hooting sounds.

Can you **MAKE THE SOUNDS** of a side-striped jackal?

DUNG BEETLE

This little critter is the world's strongest insect!

Dung beetles are very important members of the animal cleanup crew. They eat the poop, or dung, of large animals.

By **ROLLING ITS BALL OF DUNG** a few yards away from a dung pile, the dung beetle keeps its food away from other beetles.

Dung beetles **LAY EGGS in BALLS OF DUNG.** When the eggs hatch, the young have instant food.

There are many different kids of dung beetles. One kind uses starlight to look for fresh dung at night. It shapes a piece of dung into a ball. Then it uses its legs to roll the ball to a hole it has dug to bury the balls. Once the beetle has all it wants, it is time for a feast!

HOLE FOR A DUNG BALL

FACTS

KIND OF ANIMAL insect

HOME savannas in southern Africa

SIZE would fit on your little finger

FOOD dung

SOUNDS buzzing

BABIES one egg laid in each dung ball

Can you think of anything you play with that you **ROLL INTO BALLS?**

MAP OF AFRICA

This map shows where in Africa the animals in this book can be found.

RAINFOREST

Aardvark
African buffalo
African fish eagle
Chimpanzee
Diana monkey
Dung beetle
Hippopotamus
Marabou stork
Picasso bug
Southern African rock python

SAHARA

Dung beetle
Nile crocodile

AFRICA

SAVANNA

Aardvark
African buffalo
African fish eagle
African savanna elephant
African wild dog
Banded rubber frog
Black rhino
Brown hyena
Chimpanzee
Common egg-eating snake
Common warthog
Dung beetle
Giant plated lizard
Giraffe
Gray crowned crane
Hippopotamus
Impala
King baboon tarantula
Large-spotted genet
Lion
Marabou stork
Ostrich
Picasso bug
Plains zebra
Secretary bird
Senegal bushbaby
Side-striped jackal
Southern African rock python
Wahlberg's epauletted fruit bat
White-backed vulture
White rhino
Zorilla

WETLANDS

Aardvark
African fish eagle
African savanna elephant
African wild dog
Banded rubber frog
Black rhino
Common egg-eating snake
Dung beetle
Gray crowned crane
Hippopotamus
Lion
Marabou stork
Secretary bird
Senegal bushbaby
Side-striped jackal
Southern African rock python
White-backed vulture

SUB-SAHARAN DESERT

Aardvark
African fish eagle
African wild dog
Black rhino
Brown hyena
Common egg-eating snake
Common warthog
Dung beetle
Giraffe
Lion
Marabou stork
Ostrich
Picasso bug
Secretary bird
White-backed vulture
White rhino
Zorilla

PARENT TIPS

Extend your child's experience beyond the pages of this book. Visit a zoo to see in person some of the animals in the book, and to also look for other African animals. Enjoy nature shows together that feature African animals in action. Here are some other activities you can do with National Geographic's *Little Kids First Big Book of African Animals.*

WHAT'S FOR DINNER?
(PLANNING, COOKING, COMPARING)

Help your child plan and prepare a meal similar to one that animals in this book would eat. First, have them choose the chapter they want to use. Discuss your child's reasons for picking the one they did. Ask them to describe the items they would like to include in the meal that relate to the foods a carnivore, omnivore, or herbivore would enjoy. Make a grocery list together and help your child pick out items when you go shopping. Then prepare the meal and dish it up!

MAKE A BOOK
(STORYTELLING, ART)

Ask your child to pick a favorite animal in this book. Have them make up a story with that animal as the main character. Help your child write and illustrate the story on pieces of paper that you will staple or otherwise put together like a book. Have your child draw pictures to match the story they tell. Encourage them to read the story to someone else.

TAKE A HIKE
(EXERCISE, OBSERVATION, IDENTIFICATION)

Take a walk around the neighborhood or in a park with your child. Have them make a checklist before you go. List the animal categories from fact boxes in the book: mammal, bird, reptile, amphibian, and insect. On your walk, ask your child to look for examples of each kind of animal. They can write down the name of each one under the correct category. For example, a squirrel goes in the mammal section; a robin goes in the bird section. At the end of your walk, have your child count how many animals of each type they saw.

MATCH THE MEAL
(CATEGORIZING)

Help your child make a deck of cards for a matching game. Use index cards of any size for the deck. Have your child pick four animals each from chapters 2, 3, 4, and 5 of this book. They can draw a picture of each of the 16 selected animals or label each card with the name of the animal. Then have your child draw or write the name of each of these habitats on individual sheets of paper: rainforest, savanna, Sahara, sub-Saharan desert, and wetlands. Help your child spread out the 16 animal cards and line up the five habitat papers. Take turns moving one animal card at a time to the habitat where the animal lives. If an animal lives in more than one place, take turns moving it to each habitat it lives in. If your child is stuck, or wants to check their work, check the map on page 120.

GLOSSARY

AMPHIBIANS: a group of cold-blooded animals with backbones (vertebrates); the larval young of some species live in water and breathe through gills; includes frogs, toads, and salamanders

ARACHNIDS: a group of animals with no backbone (invertebrates), two body segments, and two to four pairs of legs; includes spiders, scorpions, mites, and ticks

BIRDS: a group of warm-blooded vertebrate animals that have feathers and wings and lay eggs; most can fly

BIRDS OF PREY: birds that eat meat

BURROW: a hole or tunnel in the ground that an animal can live in

CARCASS: the body of a dead animal

CARRION: the meat of a dead animal

CONTINENT: one of seven big areas of land on Earth: Asia, Africa, North America, South America, Antarctica, Europe, and Australia

DIET: what an animal eats

FUNGI (plural of fungus): living things that are not plants, animals, or bacteria; includes mold, mildew, yeast, and mushrooms

HABITAT: the home of an animal or plant

HERD: a large group of mammals, generally of the same species and usually those with hooves

INSECTS: a group of small invertebrate animals with three body segments, one pair of antennae, and three pairs of legs; many have wings

INVERTEBRATES: animals that do not have a spinal column, or backbone

MAMMALS: a group of vertebrate animals, including humans, that are warm-blooded, breathe air, have hair, and nurse their young

MATE: the male or female animal in a pair that have young

NOCTURNAL: active at night

PREDATOR: an animal that hunts for and eats other animals

PREY: an animal that a predator hunts for food

REPTILES: a group of vertebrate animals that are cold-blooded and usually slither (such as snakes) or walk on short legs (such as lizards); generally covered with scales or horny plates

RODENTS: a group of mammals with front teeth that continue to grow; includes mice, rats, and squirrels

SAHARA: Earth's biggest hot desert, which stretches across a large portion of northern Africa

SPECIES: a type, or unique kind, of animal or plant

SUB-SAHARAN: the area of Africa south of the Sahara, which is home to thousands of animals and a variety of habitats

THORAX: the part of an insect's body between the head and abdomen where the wings and legs are attached

VERTEBRATES: animals that have a spinal column, or backbone

WINGSPAN: the measure of spread-out wings from the tip of one to the tip of the other

ADDITIONAL RESOURCES

BOOKS

Donohue, Moira Rose. *Little Kids First Big Book of the Rain Forest*. National Geographic Kids Books, 2018.

Hughes, Catherine D. *Little Kids First Big Book of Animals*. National Geographic Kids Books, 2011.

Hughes, Catherine D. *Little Kids First Big Book of Birds*. National Geographic Kids Books, 2016.

Hughes, Catherine D. *Little Kids First Big Book of Reptiles and Amphibians*. National Geographic Kids Books, 2020.

Nat Geo Wild Animal Atlas. National Geographic Kids Books, 2010.

WEBSITES

A note for parents and teachers: For more information on this topic, you can visit these websites with your young readers.

animaldiversity.org

natgeokids.com/animals

birdsoftheworld.org

INDEX

Boldface indicates illustrations.

A

Aardvarks 38–41, **38–41**
Activities 122–123
African buffalo 16, **17,** 74–75, **74–75**
African fish eagles 28–29, **28–29**
African golden cats **19**
African hawk-eagles **31**
African savanna elephants 64–65, **64–65**
African wild dogs **14,** 22–25, **22–25**
African wildcats **19**
Amphibians 44, 46–47, **46–47,** 48, 124
Antelope
  impalas 16, 60–61, **60–61**
  photo gallery **62–63**
  wildebeests 16, 22, 32, **63**
Ants 38–39, 46, 87
Apes **1, 82,** 86–87, **86–89**
Arachnids 124

B

Banded rubber frogs 46–47, **46–47**
Bat-eared foxes 50–51, **50–51**
Bats 68–69, **68–69**
Bearded vultures **109**
Beetles **78, 79,** 118–119, **118–119**
Birds
  eagles 12, 28–31, **28–31**
  glossary 124
  gray crowned cranes 94–97, **94–97**
  marabou storks **104–105,** 114–115, **114–115**
  ostriches 13, **13,** 84–85, **84–85**
  oxpecker birds 81, **81**
  secretary birds **36,** 44–45, **44–45**
  vultures 13, **13,** 106–108, **106–109, 115**
Black-footed cats **21**
Black rhinoceroses 72–73, **72–73**
Blue wildebeests **63**
Bonobos **88**
Bontebok **63**
Brown hyenas 110–113, **110–113**
Brush jewel beetles **78**
Buffalo 16, **17,** 74–75, **74–75**
Bugs *see* Insects
Burrows **40,** 40–41, 124
Bushbabies 92–93, **92–93**

C

Cape vultures **109**
Caracals **21**
Carcass 124
Carnivores 12, **12,** 14–53
Carrion 106, 124
Cats, wild
  black-footed cats **21**
  cheetahs **15, 20,** 57, 111
  jungle cats **21**
  leopards **18,** 111
  lions 12, **12,** 16–17, **16–17,** 57, **57,** 85
  photo gallery **18–19**
Cheetahs **15, 20,** 57, 111
Chimpanzees **82,** 86–87, **86–87**
Cinereous vultures **108**
Collared mangabeys **101**
Common egg-eating snakes 42–43, **42–43**
Common patas monkeys **100**
Common warthogs 70–71, **70–71**
Continents 9, 124
Cranes 94–97, **94–97**
Crocodiles 12, **14,** 32–35, **32–35**
Crowned eagles **31**

D

De Brazza's monkeys **100**
Deserts 11, **11,** 111
Diana monkeys 98–99, **98–99**
Diets 12, 124
Dogs, wild **14,** 22–25, **22–25**
Dung beetles 118–119, **118–119**

E

Eagles 12, 28–31, **28–31**
Egyptian vultures **108**
Elephants 12, **54–55,** 64–65, **64–65**

F

Foxes 50–51, **50–51**
Frogs 46–47, **46–47**
Fruit bats 68–69, **68–69**
Fungi 124

G

Galagos 92–93, **92–93**
Geladas **101**
Gemsbok **62**
Genets 102–103, **102–103**
Giant plated lizards 90–91, **90–91**
Giraffes 12, **54,** 66–67, **66–67**
Gorillas **1, 88–89**
Gray crowned cranes 94–97, **94–97**
Greater kudus **63**
Grevy's zebras 58, **58**

H

Habitats 10–11, **10–11,** 124
Herbivores 12, **12,** 54–81
Herd 124
Hippopotamuses 80–81, **80–81**
Hyenas 23, 110–113, **110–113**

I

Impalas 16, 60–61, **60–61**
Insects 76–77, **76–79,** 118–119, **118–119,** 124
Invertebrates 124

J

Jackals 111, 116–117, **116–117**
Jewel bugs 77
Jungle cats **21**

K

King baboon tarantulas 52–53, **52–53**
Kirk's dik-diks **62**
Kudus **63**

L

Lappet-faced vultures **109**
Large-spotted genets 102–103, **102–103**
Leopards **18,** 111
Lions 12, **12,** 16–17, **16–17,** 57, **57,** 85
Lizards 50, 90–91, **90–91**
Locusts **79**
Long-crested eagles **30**

M

Mammals 124
Mandrills **101**
Mangabeys **101**
Mantises **78**
Map of Africa 120–121
Marabou storks **104–105,** 114–115, **114–115**
Martial eagles **30–31**
Masai giraffes **67**
Mate 124
Metallic shield bugs 77
Monkeys 13, 27, 98–99, **98–101**
Mountain gorillas **1, 88–89**
Mountain zebras 59, **59**

N

Nile crocodiles 32–35, **32–35**
Nocturnal 124
Northern giraffes **67**
Nyala **62**

O

Omnivores 13, **13,** 82–103
Ostriches 13, **13,** 84–85, **84–85**
Oxpecker birds 81, **81**

P

Parent tips 122–123
Picasso bugs 76–77, **76–77**
Plains zebras 56–57, **56–59**
Predators 12, 13, 14–53, 124
Prey 12, 125
Purple jewel beetles **79**
Pythons 26–27, **26–27**

R

Rainbow milkweed locusts **79**
Rainbow shield bugs **79**
Rainforests 10, **10,** 120–121
Reptiles
  crocodiles 12, **14,** 32–35, **32–35**
  glossary 125
  lizards 50, 90–91, **90–91**
  snakes 26–27, **26–27,** 42–43, **42–43**
Reticulated giraffes **67**
Rhinoceroses 72–73, **72–73**
Rock pythons 26–27, **26–27**
Rodents 125
Rüppell's griffon vultures **108**

S

Sahara 11, **11,** 120, 125
Sand cats **21**
Savanna 11, **11,** 120–121
Savanna elephants 64–65, **64–65**
Scavengers 13, **13,** 104–119
Secretary birds **36,** 44–45, **44–45**
Senegal bushbabies 92–93, **92–93**
Servals **19**
Side-striped jackals 116–117, **116–117**
Snakes 26–27, **26–27,** 42–43, **42–43,** 48
Southern African rock pythons 26–27, **26–27**
Southern giraffes **67**
Species 125
Spiders 52–53, **52–53,** 53
Spiny flower mantises **78**
Sub-Saharan Africa 11, 120–121, 125

T

Tarantulas 52–53, **52–53**
Termites 38–39, 46, 50, 51, 87
Thorax 125

V

Vertebrates 125
Vultures 13, **13,** 106–107, **106–109, 115**

W

Wahlberg's epauletted fruit bats 68–69, **68–69**
Warthogs 22, 27, **54,** 70–71, **70–71**
Western lowland gorillas **89**
Wetlands 120–121
White-backed vultures 106–107, **106–107**
White rhinoceroses 73, **73**
Wild dogs **14,** 22–25, **22–25**
Wildebeests 16, 22, 32, **63**
Wingspan 125

Z

Zebras
  photo gallery 58–59, **58–59**
  plains zebras 56–57, **56–59**
  as prey 12, **12,** 16, 32, 34, **34,** 35
Zorillas 48–49, **48–49**

PHOTO CREDITS

AD=Adobe Stock; AL=Alamy Stock Photo; GI=Getty Images; MP=Minden Pictures; NGIC=National Geographic Image Collection; NPL=Nature Picture Library; SS=Shutterstock

Front Cover: (elephant), Patryk Kosmider/AD; (bird), photogallet/AD; (giraffe), mattiaath/AD; (leopard), Eric Isselée/AD; (hippo), Eric Isselée/AD; (lion), Eric Isselée/AD; (background), Talva/AD; **Spine:** Eric Isselée/AD; **Back Cover:** (snake), bennytrapp/AD; (zebra), Eric Isselée/AD; (crocodile), fivespots/AD; (ostrich), fotoparus/AD; **Front Matter:** (backgrounds throughout), Talva/AD; (backgrounds throughout), Kues/SS; 1, LMspencer/AD; 2 (chimp), Ronan Donovan/NGIC; 2 (giraffe), JonoErasmus/AD; 2 (cheetah), Brian/AD; 2 (rhinos), Quintus Strauss/SS; 2-3 (zebra), Four Oaks/SS; 3 (eagle), vaclav/AD; 3 (hawk-eagle), Jo/AD; 3 (lizard), Jurgens Potgieter/SS; 3 (lion), Michael Potter11/SS; 4 (frog), Wildscotphotos/AL; 5 (eagle), Jaco Wiid/SS; 5 (hawk-eagle), Jo/AD; **Chapter 1:** 8-9, SouthernCrx/SS; 9 (UP LE), Jo/AD; 9 (UP RT), Christian Ziegler/NGIC; 9 (CTR), Alberto Carrera/AL; 9 (LO), Beverly Joubert/NGIC; 10, Fabian/AD; 11, kjekol/AD; 11 (inset), Michel/AD; 12 (UP), Ilan Horn/SS; 12 (LO), Bob/AD; 13 (LE), art_zzz/AD; 13 (RT), Martin Pelanek/SS; **Chapter 2:** 14 (UP), Jaco Wiid/SS; 14 (LO LE), bayazed/AD; 14 (LO RT), David Havel/SS; 14-15, Klein & Hubert/NPL; 16, Ann & Steve Toon/NPL; 17 (UP), oNabby/SS; 17 (LO), Nick Garbutt/NPL; 18, Sergey Gorshkov/MP; 19 (UP LE), Ludwig/AD; 19 (RT), Elana Erasmus/SS; 19 (LO LE), Art Wolfe/Science Source; 20, stuporter/AD; 21 (UP LE), Edwin Giesbers/NPL; 21 (UP RT), Frans Lanting/NGIC; 21 (LO LE), beataaldridge/AD; 21 (LO RT), davemhuntphotography/SS; 22-23, Ondrej Prosicky/SS; 23, Alex Dissanayake/GI; 24-25, Roger de la Harpe/AD; 25, Jami Tarris/GI; 26 (UP), Anthony Bannister/Avalon.red/AL; 26 (LO), Simon Hosking/FLPA/AL; 27, Alberto Carrera/AL; 28, Fotofeeling/Westend61 GmbH/AL; 29 (UP), Kiki Dohmeier/SS; 29 (LO), Stephen Lew/SS; 30-31, Theodore Mattas/SS; 30 (LO), Neil Bowman/FLPA/MP; 31 (UP), jkc916/AD; 31 (LO), Michelle Niemand/SS; 32, diegooscar01/SS; 33 (UP), Catchlight Lens/SS; 33 (LO), Anup Shah/NPL; 34-35, Mari/AD; 34 (inset), Suzi Eszterhas/NPL; 35 (inset), TJ Rich/NPL; **Chapter 3:** 36-37, Erwin Niemand/SS; 37 (UP), Suzi Eszterhas/MP; 37 (CTR), imageBROKER/M. Dobiey/AL; 37 (LO LE), Wildscotphotos/AL; 37 (LO RT), Audrey Snider-Bell/SS; 38, Peter Chadwick/Science Source; 39 (UP), Anthony Bannister/Avalon.red/AL; 39 (LO), Hope Ryden/NGIC; 40 (UP), Anthony Bannister/Avalon.red/AL; 40 (LO), Mark Jones/MP; 41 (UP), Frank Rumpenhorst/picture alliance/GI; 41 (LO), Martin Harvey/Image Source Limited/AL; 42, Anthony Bannister/Avalon.red/AL; 43 (UP), Nick Greaves/AL; 43 (LO), Tony Phelps/NPL; 44-45, phototrip.cz/AD; 45 (UP), Tui De Roy/MP; 45 (LO), Barbara Ash/AL; 46, Pete Oxford/MP; 47 (UP), David Shale/NPL; 47 (LO), Eugene Troskie/SS; 48-49, Sylvain Cordier/NPL; 48 (inset), Cordier Sylvain/hemis.fr/AL; 50, Cathy Withers-Clarke/AD; 51 (UP), Suzi Eszterhas/MP; 51 (LO), Cathy Withers-Clarke/AD; 52, Audrey Snider-Bell/SS; 53 (UP), Jabruson/NPL; 53 (LO), Photoshot/Avalon.red/AL; **Chapter 4:** 54 (UP), Sean Crane/MP; 54 (CTR), David Hosking/MP; 54 (LO), Wim/AD; 54-55, Charlie Hamilton-James/NPL; 56, Theo Allofs/MP; 57 (UP), MattiaATH/SS; 57 (LO), Andy Rouse/NPL; 58 (UP), tr3gin/SS; 58 (LO LE), phototrip.cz/AD; 58 (LO RT), henk bogaard/AD; 59 (UP LE), Jesus/AD; 59 (CTR RT), Louis/AD; 59 (LO), Phillip du Plessis/AD; 60 (UP), Daniel Y Smith/SS; 60 (LO), Ronald S Phillips/AL; 61 (UP), Richard Du Toit/MP; 61 (LO), Sergey Gorshkov/MP; 62 (LE), Jonathan Pledger/SS; 62 (UP RT), Mari/AD; 62 (LO RT), Anna/AD; 63 (UP LE), Lennjo/AD; 63 (UP RT), Ondrej Prosicky/SS; 63 (LO), Beverly Joubert/NGIC; 64-65, slowmotiongli/SS; 66, 25ehaag6/AD; 67 (LE), Windsor/AD; 67 (UP LE), Vladimir Wrangel/SS; 67 (UP RT), Marek Rybar/SS; 67 (LO LE), Zimdeck/SS; 67 (LO RT), Jen Watson/SS; 68, Doug McCutcheon/AL; 68-69, Merlin Tuttle/Science Source; 69, Merlin Tuttle/Science Source; 70, Jonathan Pledger/SS; 71 (UP), Okyela/SS; 71 (LO), Alta Oosthuizen/AD; 72, Will Burrard-Lucas/NPL; 73 (UP), maggymeyer/AD; 73 (LO), Pedro Bigeriego/AD; 74-75, Theo Allofs/MP; 75, Klaus Nigge/NGIC; 76, A. Kehinde/SS; 77 (UP), André De Kesel; 77 (LO), Daniel Rosengren; 78 (UP LE), Eugene Troskie/SS; 78 (UP RT), Andre Joubert/AL; 78 (LO), Riadi/AD; 79 (UP), noorhaswan/AD; 79 (CTR), Lorraine Bennery/NPL; 79 (LO), bennytrapp/AD; 80, Rudi Hulshof/SS; 80 (inset), camerawithlegs/AD; 81 (UP), Danita Delimont/AD; 81 (LO), melissamn/SS; **Chapter 5:** 82-83, Anup Shah/MP; 83 (UP), Ondrej Prosicky/SS; 83 (CTR), Miroslav/AD; 83 (LO), Jurgens Potgieter/SS; 84, Jason Edwards/NGIC; 85 (UP), Anton Tolmachov/AD; 85 (LO), Guy Bryant/AD; 86 (LO), Gerry Ellis/MP; 86-87, Anup Shah/NPL; 87 (UP), Anup Shah/MP; 88-89, Thomas Marent/MP; 88 (inset), Christian Ziegler/NGIC; 89 (inset), Abeselom Zerit/SS; 90, Robert Harding Picture Library/NGIC; 91 (UP), Bert Ooms/Nature in Stock/AL; 91 (LO), Fearless on Four Wheels/GI; 92-93, Des & Jen Bartlett/NGIC; 93 (UP), Stephen Dalton/MP; 93 (LO), Michal Sloviak/SS; 94, Petr Šimon/AD; 95 (UP), Anup Shah/NPL; 95 (LO), Peter Waechtershaeuser/MP; 96 (UP), Paul Hobson/MP; 96 (LO), Cheryl-Samantha Owen/NPL; 97 (UP), andreanita/AD; 97 (LO), imageBROKER/SS; 98, Edwin Butter/AD; 99 (UP), veroja/AD; 99 (LO), Vladislav T. Jirousek/SS; 100 (UP), Rod Williams/NPL; 100 (LO), Tomas Drahos/SS; 100-101, Thomas Marent/MP; 101 (UP RT), Juergen & Christine Sohns/MP; 101 (LO), Thomas Marent/MP; 102, Ann & Steve Toon/NPL; 103 (UP), Ariadne Van Zandbergen/AL; 103 (LO), Miroslav/AD; **Chapter 6:** 104 (LE), Philip Marazzi/Papilio/AL; 104 (RT), henk bogaard/AD; 104-105, Matyas Rehak/AD; 106, Jeff Huth/AD; 106 (inset), 2630ben/AD; 107 (UP), Lennjo/AD; 107 (LO), slowmotiongli/SS; 108 (UP LE), Sourabh Bharti/SS; 108 (UP RT), Cinematographer/AD; 108 (LO), Jag Images/AD; 109 (UP), Eleanor Esterhuizen/SS; 109 (LO LE), petrsalinger/AD; 109 (LO RT), phototrip.cz/AD; 110, EcoView/AD; 111 (UP), Bill Gozansky/AL; 111 (LO), Martin Harvey/Biosphoto; 112-113, Suzi Eszterhas/MP; 113 (inset), Solvin Zankl/NPL; 114, Borkowska Trippin/SS; 115 (UP), Alex254/Wirestock Creators/AD; 115 (CTR), Marek R. Swadzba/AD; 115 (LO), Gabrielle/AD; 116, Chris Fourie/AD; 117 (UP), Beverly Joubert/NGIC; 117 (LO), Gallo Images/GI; 118-119, henk bogaard/AD; 119, Mipa Photo/AD; **Back Matter:** 122, Lumos sp/AD; 123 (UP), Regalado Santos/SS; 123 (LO), pixelrobot/AD; 123 (drawing), JJ Chamon/AD; 128, Tomas Hulik/AD

FOR MY YOUNGEST GREAT-NEPHEWS, WESTON AND ELIJAH, BORN AS THIS BOOK WAS WRITTEN; AND FOR MY FAVORITE TRIO OF YOUNG READERS, NEIGHBORS BERUH, AMAN, AND DAGEM —C.D.H.

Since 1888, the National Geographic Society has funded more than 14,000 research, conservation, education, and storytelling projects around the world. National Geographic Partners distributes a portion of the funds it receives from your purchase to National Geographic Society to support programs including the conservation of animals and their habitats. To learn more, visit natgeo.com/info.

For more information, visit nationalgeographic.com, call 1-877-873-6846, or write to the following address:

National Geographic Partners, LLC
1145 17th Street NW
Washington, DC 20036-4688 U.S.A.

More for kids from National Geographic: natgeokids.com

National Geographic Kids magazine inspires children to explore their world with fun yet educational articles on animals, science, nature, and more. Using fresh storytelling and amazing photography, *Nat Geo Kids* shows kids ages 6 to 14 the fascinating truth about the world—and why they should care. natgeo.com/subscribe

For rights or permissions inquiries, please contact National Geographic Books Subsidiary Rights: bookrights@natgeo.com

Designed by Brett Challos

Library of Congress Cataloging-in-Publication Data

Names: Hughes, Catherine D., author.
Title: Little kids first big book of African animals / Catherine Hughes.
Other titles: National geographic little kids.
Description: Washington, DC : National Geographic Kids, 2025. | Series: National Geographic kids books | Includes index. | Audience: Ages 4-8 | Audience: Grades K-1
Identifiers: LCCN 2024010832 | ISBN 9781426374005 (hardcover) | ISBN 9781426375484 (library binding)
Subjects: LCSH: Animals--Africa--Juvenile literature.
Classification: LCC QL336 .H83 2025 | DDC 591.96--dc23/eng/20240708
LC record available at https://lccn.loc.gov/2024010832

The publisher would like to thank herpetologist and National Geographic Explorer Ruchira Somaweera, bird expert Jonathan Alderfer, zoologist Lucy Spelman, and entomologist William Lamp for their expert review of this book. Thanks also to Grace Hill Smith, project manager, and Sharon Thompson, researcher. Book team: Emily Fego, project editor; Lori Epstein, photo manager; Colin Wheeler, photo editor; Mike McNey, map production; Alix Inchausti, senior production editor; Yogi Carroll and Lisa Walker, production managers; and David Marvin, associate designer.

Printed in China
24/RRDH/1